Pan-African Futurism

Pan-African Futurism

Ghana and the Paradox of Technology for Development

Reginold A. Royston

UNIVERSITY OF CALIFORNIA PRESS

University of California Press
Oakland, California

© 2025 by Reginold A. Royston

Cataloging-in-Publication data is on file at the Library of Congress.
ISBN 978-0-520-42341-1 (cloth
ISBN 978-0-520-42342-8 (pbk.
ISBN 978-0-520-42343-5 (ebook)

GPSR Authorized Representative: Easy Access System Europe,
Mustamäe tee 50, 10621 Tallinn, Estonia, gpsr.requests@easproject.com

34 33 32 31 30 29 28 27 26 25
10 9 8 7 6 5 4 3 2 1

This book is dedicated to the ancestors,
Carmen Mitchell, Emmanuel Okyere,
Akua Akyaa Nkrumah, Janet Kwami,
Herman Chinery-Hesse, and all future
Pan-Africanists.

Contents

Acknowledgments

A vast web of comrades, compadres, and Ghanafuo helped me to deliver this book, almost two decades after entering graduate school. I am humbly indebted for their patience, support, and wisdom to help me see beyond the horizon.

This research and book would not have been possible without the support of my Ghana family, Duke Ofori, Erin Ransom, and their children, Kareem Bobtoya, Christa Sanders, my Cali-brother Kwesi Wilson, Panji Anoff, Nii Oblie Ardayfio, Louisa Addio, and so many more. I am grateful for the acknowledgment and support of my research in the greater Ghanaian diaspora, especially to the Ghanaian National Council of Chicago and the Ghanaian National Council of Northern California. Thanks to my colleagues at the University of Ghana, Legon, who have championed my work, including Akosua Adomako Ampofo, Akosua Darkwah, Nana Akua Anyidoho, Mjiba Frehiwot, Ọbádélé Kambon, Audrey Gadzekpo, Amos Anyimadu, and so many others. Thank you to friends at Ashesi University, including Joseph Oduro-Frimpong, Ayorkor Korsah, Nathan Amanquah, and Kobby Graham. The members of GhanaThink have been important research contacts, interlocutors, and a source of inspiration for this book, including Ato Ulzen-Appiah, Nehemiah Attigs, Gameli Adzaho, Fiifi Baidoo, Jemila Abdulai, Patrick Keli Atitsogbui, Dominic Kornu, Kola Nut, and Eryam Tawia, as well as members of the BloggingGhana community such as Edward Amartey-Tagoe and Dr. Kajsa Hallberg Adu, among many others.

This book had two very strong intellectual incubators: the University of California, Berkeley, and the University of Wisconsin, Madison, two transformational public institutions that have foundationally supported Black studies, African studies, cultural studies, and new media and have helped me make indelible links back to the continent. At Berkeley in particular, I met Abena Osseo-Asare, who helped me to see Ghana, as I worked as an editor on the *Atomic Junction* documentary. This book project began as research with Jenna Burrell, whose insight and guidance I am forever indebted to. Percy Hintzen, ever the provocateur, suggested that my inquiries into Black life and digital media should extend beyond the diaspora and back to Africa; the Department of African American Studies at Berkeley made this happen under the guidance of Michel Laguerre, Ugo Nwokeji, and the late, great Robert Allen, and through funding from the Vèvè Clark fund. Using this, I was able to travel to Ghana for the first time in 2011, and I am forever indebted to my mentors and friends Leigh Raiford, Na'ilah Nasir Suad, Ula Taylor, Heike Raphael-Hernandez, and many others, especially Martha Saavedra at the Center for African Studies. The Berkeley Center for New Media was also a generous benefactor and home for me, and I am forever grateful to Greg Niemeyer, Ken Goldberg, Keith Feldman, and especially Abigail de Kosnik for their inspiration and support over the years. I am indebted as well to the work and guidance of Mano Delea and Kwame Nimako in doing research in Amsterdam.

The University of Wisconsin, Madison, has vitally nurtured this book project and my development as a scholar, and I am grateful to my current and former colleagues in the Department of African Cultural Studies, including Aliko Songolo, Luís Madureira, Marissa Moorman, Adeola Agoke, Matthew Brown, Jacqueline-Bethel Tchouta Mougoué, Ainehi Edoro, Vlad Dima, Sam England, K. D. Thompson, Nevine El-Nossery, Damon Sajnani, Mustafa Mustafa, Warrick Moses, Elaine Sullivan, Bill Bach, and Toni Landis. Thanks to the tireless and crucial support of Aleia McCord at the UW African Studies Program and my fellow Wisconsin Africanists Emilie Songolo, Neil Kodesh, Gregg Mitman, and so many others. Most of all, thanks to Teju Olaniyan, whose words have been with me during the entire writing process.

In the Information School, this work would not have been possible without the enthusiastic support of Kristin Eschenfelder, Alan Rubel, Catherine Arnott Smith, Sunny Kim, Michele Besant, Greg Downey, Rebekah Willett, Jacob Thebault-Spieker, Corey Jackson, Tracy Lewis Williams, Dorothea Salo, DeAnza Williams, Megan Adams, Anjali Bhasin, and especially Jonathan Senchyne, as well as our amazing staff and administrators through the years. The Robert and Jean Holtz Center for Science and Technology Studies has proved an indispensable intellectual community for me, thanks to Noah Weeth Feinstein, Jenell Johnson, Nicole Nelson, and Sainath Suryanarayanan. Special thank yous to my campus academic circle for their inspiration and support, including Ethelene Whitmire, Christy Pujara-Clark, Michael Thornton, Sandra Adell, Thulani Davis, Anthony Black, Mosi Ifatunji, and many others in African American studies; Jeremy Morris and Lori Lopez in communication arts; Erica Turner, Linn Posey, Keisha Lindsay, Gideon Amoah, Mary Beth Collins, Raphaëlle Rabanes, Jerome Camal, Nancy Kendall, and Michael and Rachelle Winkle-Wagner, Amy and David Washbush Hilgendorf and the Linden Cohousing community, among dozens of others.

Many thanks to Catherine Cole and James Yékú for reading early drafts of this work through the UW Center for Humanities First Book Award program. Thanks also to Victoria Bernal for early feedback and strengthening my analysis, and to Tom Boellstorf for his mentorship via the Social Science Research Council. The Summer Collaborative for African Languages Institute (SCALI) at Michigan State University helped start my journey in the Akan-Twi language, and I am grateful to F. Asare Kena, Kwame Assenyoh, and Rachel Flamenbaum for my time there. Special thanks also go to Ouseina Alidou and Alamin Mazuri at Rutgers University for helping to spread my work. *Medaase* goes to my *ebusua* at the Ghana Studies Association and the greater African Studies Association (USA). Abdul Alkalimat, André Brock, Meredith Clark, and Adam Banks were important interlocutors and have helped me to center my work in Black studies over the years. Special thanks to Neil Roberts, Seulghee Lee, Vivian Huang, Maurita Poole, and my neighbor Ethan Zuckerman during my time in the Berkshires; and to Jerome Taylor (RIP), Michele Reid-Vazquez, and Yolanda Covington-Ward at the University of Pittsburgh. I drew inspiration from Faisal Abdu'Allah and the Black Arts and Data Futures workshop and my colleagues at AI for Afrika, especially S. Ama Wray, José Cossa, Femi Omere, Ash Baccus Clark, Alex Tsado, and many others.

I am indebted to and humbled by the earnest assistance I have received from my students and on-the-ground collaborators, without whom this work would never have been so done well, including Fauziyatu Moro, Theodora Eyram Amaglo, Sera Kevuti, Charlotte Von De Bur, Yinka Ajibola, Vincent Ogoti, Tolulope Adelabu, Hiwot Adilow, Kevin Bannerman-Hutchful, Arielle Steele, Ebenezer Tetteh, and Sheriff Issaka.

Special thanks go to students in my seminars African and the Internet, Pan-African Returns, Orality and Technology in Africa, and Technology and Development in the Global South. Thanks for your vision, Michelle Lipinski and Jyoti Arvey at University of California Press, and to the anonymous reviewers, designers, and editors; and to Chris Lura early on.

The work on this book would not have been possible without my partners in the beautiful struggle of this academic life, including thought-partner Krystal Strong and dedicated colleagues in this field such as Ben Talton, Seyram Avle, Mario Nisbett, Justin Gomer, Amir Hasan Loggins, Jasminder Kaur, Ndirangu Wachanga, Msia Kibona Clark, Roderick Carey, Irene Chen, Caitlin Marshall, Andrew Godbehere, Ashley Ferro-Murray, David Humphrey, Kris Fallon, Alenda Cheng, Margaret Rhee, Ronald Williams II, and Charisse Burden-Stelly. Shout out to all my academic families and their children, to name a few, Harry and Clementine Odamtten, Dotun and Yemi Ayobade, Sepehr and Kihana Ross-Vakil, Paul and Elisheva Cherlin, Zakee Sabree and Lauren Jones, Natalie Moore and Rod Falls, and Ron and Thai-An Kimmons-Ngo, as well as the Oakland crew, Madison crew, and the Chicago/Evanston crew. I am indebted to my longtime intellectual elders Louis E. Wright, Michael Blakey, Mark Mack (RIP), Segun Gbadegesin, Linda Heywood, Sylvia Wynter, and Louis E. Wilson (RIP).

I am most grateful for the love and support of my family in the years it took to complete this work, especially my brothers, always ready with the jokes, my proud father Al, and my mother Irene for paving a path in academia and on the front line. I would be nowhere without my in-laws, including the ever-adventurous Alex and Roni McKinney, and my coconspirators in this #afrogeeklife—Erin Horne McKinney, Zeke, and Jerome—as well as Derek McKinney, who teleported us early on to the motherland. This work was nurtured with the love and support of many mammas and aunties, including Alexis Camille, Titilayo Makini, and Rachel Verdell. Without a doubt, praises go to my loving and fearless partner, the most rigorous of advisers, Dr. Maxine McKinney de Royston. To my three children, I hope you will find in this book diligent work that lives up to the family name.

Abbreviations

AfCFTA	African Continental Free Trade Area
AU	African Union, since 2002, the continent's chief multilateral political entity, successor to the Organization of African Unity (OAU), founded in the 1960s
ECOWAS	Economic Community of West African States, a multilateral organization
GSMA	Groupe Spécial Mobile Association, a mobile Internet service providers' trade group, initially founded to support the Global System for Mobile (GSM) cellular phone standard
ICT or IT	information and communication technology or information technologies, used inter-changeably in this text
ICTD or ICT4D	information and communication technology and development OR information and communication technology for development
ISP	Internet service provider (e.g., Vodafone or MTN)
ITU	Internation Telecommunications Union, a UN agency
KACE or AITI-KACE	Kofi Annan Centre for ICT Excellence or Ghana-India Advanced Institute for

Technology Innovation at the Kofi Annan
Centre for ICT Excellence in Accra,
a government agency that services telecom
access for Ghana and ECOWAS

NCA   National Communications Agency,
a regulatory agency in Ghana

SMS   short message service, also known as text
messaging

Introduction

Disrupting the Network Society

In 2019 Ghanaian president Nana Akuffo-Addo initiated a yearlong tourism campaign labeled "The Year of Return," enticing Afrodescendants of slavery to explore their roots on the continent. Concerts and heritage festivals were held all year long, with social media posts attesting to pilgrimages by Black American celebrities. *Essence* magazine hosted a "Full Circle" version of its lucrative annual music festival in Accra for the first time that summer. The annual number of foreign visas to Ghana rose to 800,000 in that year alone, an increase of as much as 10 percent.[1] In a spectacular moment in December, a group of more than 140 African American and Caribbean expats received citizenship in a ceremony hosted by the Ghanaian president.[2] In an effort to engender racial solidarity and increase investment from Blacks living abroad, the government-sponsored event was a spectacle of traditional robes and new names to go with new passports. "On behalf of the government and people of Ghana, I congratulate you on resuming your identity as Ghanaians. *Akwaaba* [You are welcome]," said Nana Akuffo-Addo from the Jubilee House, the country's presidential residence, which resembles an Akan royal stool.[3]

A few weeks earlier, down the road at the Accra International Conference Center, the German software firm SAP and the American social network LinkedIn were involved in a different kind of cultural exchange. Speakers gave "rapid-fire" talks on launching start-ups and entertained

FIGURE 1. Volunteers and participants watch speakers give "ignite" talks at the annual Ghana Tech Summit in 2018 in Accra. Photo by author.

pitches from aspiring entrepreneurs at the 2nd Ghana Tech Summit. Ato Ulzen-Appiah, an electrical engineer educated at MIT and Stanford, discussed his work with GhanaThink Inc., a nonprofit that's been networking software developers and young professionals for monthly meetups around the country for more than 10 years. Recalling the achievements of the Nkrumah era's Akosombo Dam and the Ghana Atomic Energy Commission, he exhorted the many students, start-ups, and entrepreneurs at the networking conference to live up to the mission of his organization's decidedly Pan-Africanist objectives: "Our goal is to build a critical mass of patriotic, passionate, positive, proactive, progressive programmers, who will build a bright future for all of Africa."[4]

. . .

Technology marks both a problem and its solution. This book describes how tech users with roots in Ghana, West Africa, deploy information and communication technology (ICT) to overcome the problem of *disjunction* between the West and the Global South. More than a digital divide in the awareness and usage of technology, Africa has been linked asymmetrically to the global Internet since its inception in the 1990s. As of 2024, the African continent had nearly one-quarter the Internet bandwidth and capacity of either Europe or North America. African countries dominate the United Nations' annual list of least connected countries, in which Internet is most expensive and access is often inconsistent.[5] Ghana, the nation known as Africa's "Black Star" in part for its economic development success and democratic stability over the past 30 years, has been a notable exception. The West African republic has been among Africa's earliest adopters, gaining access to the Transmission-Control Protocol/Internet Protocol (TCP/IP), aka the Internet, in 1995 ahead of most of the continent. Today, Ghana is among the most highly connected sub-Saharan African countries, with six landing stations for crucial undersea fiber-optic cables enabling transnational digital communication.[6] Connectivity has enabled exchange between Ghanaians, Ghanaians living in diaspora, and digital access to the broader world. Yet while mobile phone penetration rates reach well past 60 percent in this country of 30 million people, only about 40 percent of Ghanaians have access to robust Internet.[7] At one point Ghana boasted more nationwide, major Internet service providers (ISPs) than either the United States or United Kingdom, but its digital networks are overall more unstable and technically unreliable than those in the West, where its diasporans reside. This disjuncture exacerbates conditions of alienation between the homeland and diaspora and extends a system of unequal relations between Africa and the West set in motion at the dawn of colonialism. The experience of the Internet and its promises of a more democratic "network society" remain deeply uneven in Ghana, in an age when digital inaccessibility and socioeconomic inequality are indelibly linked.

The solution to this disjuncture for the media activists and software developers that I document in this book has been to not simply embrace the transformative affordances of ICT as promised by a neoliberal world economy. Rather, with a sense of *Pan-African futurism,* the developers documented in this book hold to a liberating vision of "one destiny" by connecting through digital media to advance Ghana and all Africans' technical and political autonomy. In the following chapters, I describe the work of civic-minded software engineers (*activist-developers*), alongside media producers working in diaspora (*digital diaspora*) who utilize STEM (science, technology, engineering, and math) practices as a means of increasing social and economic development steeped in a Pan-Africanist ethos.

Political historian Hakim Adi has characterized Pan-Africanism as ideas that are "concerned with the social, economic, cultural and political emancipation of African peoples, including those of the African diaspora. What underlies the manifold visions and approaches of Pan-Africanism

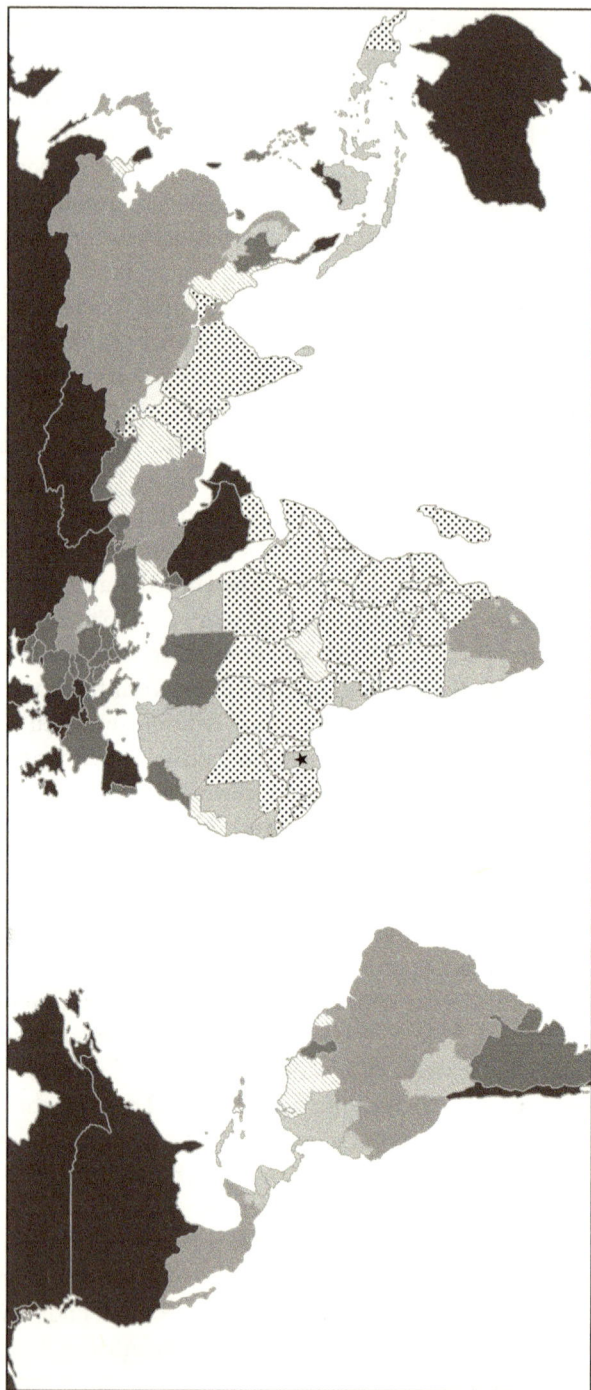

MAP 1. A map of Internet usage across the world illustrates that the African continent is the world region with the lowest rates of connectivity.
Source: International Telecommunications Union, United Nations, maps accessible at https://datahub.itu.int/.

10–44.2% ● 44.2–74.3% ● 74.3–85.1% ● 85.1–92.2% ● 92.2–100% No Data

and Pan-Africanists is a belief in the unity, common history and common purpose of the peoples of Africa and the African diaspora and the notion that their destinies are interconnected."[8] While scholars and stakeholders have debated its central tenets and practices, Pan-Africanism has been an enduring social, economic, and political sensibility embraced by diasporans and Africans on the continent. From the Garveys' Universal Negro Improvement Association (UNIA) to scholar-activist W. E. B. Du Bois and statesmen such as Léopold Senghor to revolutionary Frantz Fanon, Pan-Africanists of the early to mid-20th century held competing and at times oppositional philosophies and strategies in their advocacy of political freedom for Africans and their diasporas. However, most were united in a common vision of African continental freedom and global Black solidarity, an ethos that would lead to independence from colonial states and equal relations with the greater world.

While in the 21st century Pan-Africanist diplomatic ties and cultural circuits are enjoyed across borders, the historic movement leveraging transnational and racial unity has struggled to account for deep cleavages in legal, regional, and economic cooperation and is challenged by xenophobia or persistent illiberal governments. For example, while Nigeria and Ghana enjoy a fraternal relationship as former British colonies in West Africa, memories of tit-for-tat migrant expulsions from their respective countries during the 1970s and 1980s remain searing just beneath the surface of these relations, the basis for the colloquial phrase "Ghana Must Go."[9] Even with the multilateral success of the African Union since 2002, the Economic Community of West African States (ECOWAS), and the African Continental Free Trade Area (AfCFTA) launched in 2021, Pan-Africanism has been problematic in the postcolonial era. Pan-Africanism did little to stem regional genocide in Rwanda in 1994 (in part leading to the dissolution of the Organisation of African Unity, OAU), and has failed to de-escalate conflicts in Somalia, Sudan, and elsewhere. Ebola and COVID-19 outbreaks were mostly met with protectionism rather than collaboration in the 2000s. In 2017, when The Gambia's president Yahya Jammeh refused to step down after 22 years in office (not the longest African authoritarian regime by far), ECOWAS was first roundly criticized for its reticence to act and later critiqued for using armed troops to force a political transition, a move many perceived to be at the behest of the West. Amid these failings and many others, idealistic Africans still seek to elide regional and legal distinctions on the continent through Pan-Africanism, utilizing its many iterations to imagine Africanness unlimited by national boundaries.

Today, it is also evoked by Ghanaian entrepreneurs and social activists to advocate for market-driven, neoliberal models of national development. While the usages may seem at odds, what connects tech activists, civil organizers, and social media users to the movements of the independence era is an enduring resistance to foreign dependency and a redemptive vision for African political and cultural unity, grounded in tangible *and* material outcomes. This book documents the tacit and implicit efforts of civic activists, tech entrepreneurs, media producers, and everyday users in Ghana and its diasporas to reconfigure ICT and the unequal relations of the information age. For many, the term *Pan-African* has become a buzzword for firms and start-ups seeking to brand their business and marketing strategies across different countries on the continent. Used in these ways, *Pan-Africanism* is far removed from the political philosophy most identified in Africa with Ghana's first president, Kwame Nkrumah, an ardent socialist who advocated for "a new social synthesis in which modern technology is reconciled with human values, in which the advanced technical society is realized without the staggering social malefactions and deep schisms of capitalist industrial society."[10]

Since 2008 I have been a participant observer in Ghanaian software developer communities, starting in San Francisco, Chicago, Amsterdam, and ultimately in Accra and other urban sites in the homeland. I have participated in online spaces important to Ghanaians and in the African mediascape broadly via GhanaWeb.com, Twitter/X, hi5, Facebook, and WhatsApp. This research documents the work of innovators, including the struggling network engineers who become social advocates during elections; the flow of mobile phones and tools to and from diaspora; and the emergence of start-ups and midsize companies such as theSoftTribe and HubTel. This book synthesizes my experiences as an ethnographer, Black diasporan, and tech interlocutor who, upon observing Ghanaian tactics of innovation in this moment, is ultimately prompted to ask: *If neoliberal governance, ICT innovation, and financial globalization are proffered as the routes to economic self-sufficiency in the information age, what is the role for the notions of indigenous autonomy and freedom as originally envisioned by Pan-Africanism?*

BLACK STAR RISING

Since the early 2000s, the African continent has been experiencing an upsurge of commercial interest internally and political interest from foreigners and African return migrants alike. The publication of a 2011 cover article in *The Economist* titled "Africa Rising" seemed to give a slogan to a set of positive developments that had been stirring for a decade.[11] In the past 20 years Africa has been the fastest growing global region for mobile phone adoption and currently boasts more than 613 million mobile subscriptions, up from 250 million in 2008.[12] Vernacular tech practices, such as Kenya's M-PESA application, emerged as immensely successful text-message/money transfer services. These mobile money tools required no bank accounts and capitalized on the growing popularity of mobile phones, pioneering digital wallet transactions worldwide.[13] At the same time, Nigeria's Globacom (GLO) became one of the first African telecoms to build its own fiber-optic connection between the US and West Africa. At the time the article was published, Ghana's gross domestic product (GDP) grew to 14 percent in 2011 (up from 7.9 percent in 2010), and it was described as the world's fastest growing economy, with Nigeria also a top-10 growth country in the same period.[14] Soon after, the World Bank reclassified nations such as Kenya and Ethiopia from "developing" to "middle-income" economies, signaling recognition of a resurgent African "middle class."[15]

In the years since, India has partnered with Ghana to build dams and ICT centers in the country, and Arab states have invested in agricultural projects in East Africa, with Qatar lending millions to build a port in Kenya. China has become an influential systems builder, installing telecom networks across the continent via Huawei,[16] and it has become one of Democratic Republic of Congo's largest coltan-mining partners. In this "rising" discourse, African transnationals who returned home to set up businesses and reestablish roots were especially significant, in part due to African émigrés gaining more visibility in Western popular culture, academia, and business. The election of the first Black president of the US, Barack Obama, whose father was from Kenya, alongside historic victories by Ghana's soccer team at the first World Cup championship in South Africa for the Fédération Internationale de Football Association (FIFA) and the declaration of the International Decade of People of African Descent by the United Nations (2015–24), bolstered a sense that Africa's futures were rising. The fervor around Ghana's potential in ICT peaked in 2018 when Google made a surprise announcement that it would be establishing its next artificial intelligence research center, not in Europe or Asia—regions where AI has viscerally become a part of everyday life—but rather in Accra.[17]

In *Pan-African Futurism*, I offer a way to think about Africa's early 21st-century successes as more than the entanglement of the technological and the economic, that is, the "Digital Revolution" meets "Africa Rising." Much of the groundwork for these developments in tech has been laid by pioneering firms that have localized Western software to Ghana's needs. Ghana's vibrant tech space has also been opened up by social media–driven civil society campaigns led by GhanaDecides, GhanaThink, Pen-PlusBytes, and others working for sociopolitical and economic inclusion for youth and people typically underrepresented in its politics. Rather than simply reaping the achievements born from adopting the social cues and digital techniques of Silicon Valley "disruptors," these African digital actors also embrace an ethos of political liberation and techno-optimism, or as I describe it, a sense of *Pan-African futurism* drawing on the history of Ghana as one of Africa's first independent republics.

DEFINING PAN-AFRICAN FUTURISM

In arts and literary circles, Afrofuturists in the diaspora and Africanfuturists based on the continent have seized upon the imagery of science, technology, and sci-fi to create a speculative vision of freedom for dispossessed people of African descent. I use the term *Pan-African futurism* to describe a complementary utopian perspective, rooted in tangible practices that are already taking place on the ground, in Ghana, in diaspora, and in Africa broadly, involving tech workers, artists, entrepreneurs, civic activists, and returnees. In the process of adopting, modifying, and reinterpreting digital tools for a society beset by social and infrastructural instability, these social actors are in fact engaged in the creation of new technologies—that is, as science and technology studies (STS) theorist and sociologist Bruno Latour famously wrote, they are engaged in practices that "make society durable."[18]

Rather than simply following a narrative of tech-driven economic growth, opportunism, or even "modernization" and "development," Pan-African futurists like Ulzen-Appiah and others are motivated by ideas and tactics from anti-colonial leaders and political resistance from the late 19th and mid-20th centuries. Pan-Africanism has also been transformed in the process, from a vision of centralized political and economic authority for African governments into a more diffuse and networked form of solidarity, no longer tied to its socialist and statist past. I identify this philosophy as a driving force behind tech innovation among the interlocutors of this book, and also a key source of tension for my contacts and interviewees as they find success with technology in their development projects in Ghana. This book explores two important questions that highlight the paradox of technology for development in Africa: How do Ghanaians balance the potential for technology to be economically transformative with the justice-oriented ethos that drove Pan-Africanism in its inception? If liberation is indeed the goal, how can the development promised through collectivist ideologies from the 20th century be achieved through "social enterprise" or the wholesale embrace of market capitalism in our era for Africa's social- *and* profit-minded tech entrepreneurs?

Historically, Pan-Africanism is the belief that political unity among Africans and Africa's diasporas should drive collective action toward their greater well-being. It can be debated which movements or texts provide the starting point for an African-centered political sense of the world—often in opposition to a reterritorialization within the empires of the West—but the label Pan-Africanism has held cogent political and cultural sway since resistance to the Transatlantic Slave Trade in the 18th century and for Africans dispossessed by European colonization since the 19th century. The first "Pan-African"-themed conference was held in 1900 in London, organized by Trinidadian lawyer Henry Sylvester Williams, codifying the emerging philosophy as a movement against racism and colonialization. By the time of the fifth Pan-African Congress in 1945, Pan-Africanism had become mostly associated with dissident leaders such as Kwame Nkrumah of Ghana and organizers who would become presidents, including Jomo Kenyatta of Kenya, Nnamdi Azikiwe of Nigeria, and Hastings Banda of Malawi, as well as Caribbean statesmen and advocates such as Amy Jacques Garvey and W. E. B. Du Bois.[19]

Pan-African leaders such as Senghor, Marcus Garvey, Azikiwe, and Nkrumah were political theorists and statesmen whose careers were enveloped in the 20th-century notion of modern industrialization—factory systems and emergent supply chains defined by colonialism, world wars, and the expansion of the Western corporate enterprise. Through the rapid globalization of the economies and telecommunications of the last century, a world system was remade—and the interest of the Pan-Africanists at that time was to seek both political and economic power through the direct control of the state. Enterprise, whether political, industrial, or agricultural, was at the heart of Pan-Africanist government policies, and the newly independent states were as acutely attentive to their role as entrepreneurial managers of their new nations as to their own political futures. When Ghana achieved political independence in 1957 it took Garvey's UNIA symbol of the Black Star as a centerpiece in the new nation's flag. In choosing the icon of the Jamaican entrepreneur and nationalist's shipping service (via Garvey's *Black Star Line*), the leaders of Ghana at the time wedded the notion of progress to both a transnational movement for Black independence and a symbol of industrial innovation that was seen as instrumental in building national autonomy. The Black Star on Ghana's flag today is not only a symbol of indigenous empowerment, but also a sign of transit and return between diaspora and the homeland. As a symbol for Africa's rise and industrial power, Ghana wedded itself during independence to a vision of freedom tied to industrial innovation and agency for all African and colonized people. Ghana became in many ways the first Pan-African republic on the continent,[20] and it only seems appropriate that a sense of what I am calling Pan-African futurism pervades the progressive vision of Ghana's tech developers and social media activists today.

In the history of such futurisms, we should acknowledge precedents in the Italian futurist movement of the early 20th century, and less so in the American futurist forecasting of Alvin Toffler. Toffler's eclectic social science, born from journalistic writing on early AI and Silicon Valley, largely attracted the interests of mainstream social science and corporate strategists.[21] The Italian futurist movement (1900s–1930s) was a political and arts movement that embraced machines and industrialization in order to break with what idealogues felt was Europe's anti-modernist political economy.[22] In 1994 the White American writer Mark Dery used futurism, in a non-fascist sense, to characterize the analogies that sci-fi writers such as Samuel Delaney and Octavia Butler and electronic musicians (DJs and producers) were making between the alienation of Black subjects in the West and the "abduction" experiences of African Americans born of the Transatlantic Slave Trade. In conversation with writers such as Greg Tate, Mark Sinker, and the Ghanaian British writer Kodwo Eshun, the idea of *Afrofuturism* emerged as a discourse about Blackness and technology on the cusp of the consumer computing revolution.[23] In the 2010s the idea would be revisited by novelists such as Nnedi Okorafor using the neologism *Africanfuturism* to emphasize Africa's indigenous precedents in the fantasy genre and for experiences local to the continent, where race was not always a primary subtext.[24]

Pan-African futurism describes the ways that African innovators on the continent and in the diaspora are utilizing technology and digital media for connection and liberation in our time. By invoking the ideas of technological adoption and collective industrial advancement drawn upon by Nkrumah and others, Pan-African futurists today are endeavoring to realize a vision for a free and connected Africa, forgoing polities that historically divided communities and availing Africans of the freedom of movement and exchange that the notion of a network society promised. For my interlocutors, this social imaginary is not limited to the territory of Ghana or to the African continent, but extends to Africa's (digital) diasporas across the globe as well. Pan-Africanism represents a redemptive vision for African and Black solidarity, justice, political unity, and cultural and spiritual connections, whose foremost concern historically was the reordering of the material conditions of Africa and Afrodescendants. The focus on an eminent, free, progressive Africa, free from domination by foreigners, is central to the liberatory vision of Pan-Africanism: It has always been future oriented,[25] focused on economic and industrial goals. Today its futurism has wedded its political destiny to the capacities of digital media and ICT. As such, this is a Pan-Africanism that cannot necessarily exist without an explicit or implicit politics: I use the capitalized version of the term, rather than a more generalized geographic notion, as in *pan-Africanism*. While Rabaka and other theorists have highlighted the ways in which the informal, everyday moral choices and positions of noninstitutional African actors can be their own forms of political agency,[26] for me these (lowercase) "pan-Africanisms" cannot be divorced from an explicit political ethos of unity that pervades public, and especially racial, discourse on the continent. I do not find the distinction useful for the purposes of this book. While their work may indeed be forms of everyday, or (lowercase) pan-Africanism, the discourse and actions of my interlocutors were often explicitly Pan-African in political character, as evidenced in Ulzen-Appiah's remarks cited previously.

The practices to which I am referring are not speculative; however, Ghana's activist developers do owe some of their inspiration to the Afrofuturist sentiments that have been nibbling at the edges of its projects. The mostly diasporan Afrofuturism and indigenous-centered Africanfuturism have emerged as concurrent discourses that speak to visions and creative explorations of Black life amid the evolving entanglements and opportunities of the Internet age. As aesthetic movements, typically Pan-African in scope,[27] we cannot dismiss the role that Afro/Africanfuturism plays in the lives of Pan-African futurists. Wanuri Kahiu's 2008 eco-dystopian film *Pumzi* pioneered an Africanfuturist aesthetic in Kenyan cinema that was amplified globally with the release of the 2018 film *Black Panther* (Disney/Marvel). These were among the many cultural productions that identified an emergent movement around sci-fi fandom in a global Black media public. Yet we should critique any gestures that simply attach Black success or facility with STEM to Afro/Africanfuturism. Enthusiasm on the African continent for the possibilities of "leapfrog" ICTs was signaled early in the 21st century with descriptions of a mobile phone "revolution" in Africa. The fervor was sustained via virtual diaspora message boards, political activism via social media, and the emerging *Afropolitan mediascape* (see chapter 4). During the time of this research (2008–24), Ghana's activist developers and many other African and diaspora digital actors were heralding the role of STEM and specifically digital technology as a social leveler used to create community and push back against oppression amid episodes such as #BringBackOurGirls in Nigeria and #RhodesMustFall in South Africa.[28] While the heroic image of Wakanda has come to encapsulate that potential for many,[29] the enabling power of technology was already well established in the Afro-Arab Spring and resulting Occupy movements in places such as Nigeria, Ghana, and Senegal, beginning in Tunisia 2010.[30] Pan-African futurism in this book specifically seeks to describe the 21st-century visions for African economic and social liberation, in which the role of technology is integral. These gestures can be distinguished from the industrializing efforts of the modernization era of national development in the 1960s across the continent, as the Pan-African futurists here often work independently of the state. Not content with the space of myth and fantasy, activist-developers are realists, engaged in practices that lay the groundwork for an African socio-/techno-liberationist future. To be clear, *Pan-African futurism* is my term for the practices and discourse of the tech developers and social activists documented in this book. In Ghana, my research contacts often referred to themselves as *disruptors, innovators, social entrepreneurs,* or *patriots,* among other terms. The sense of unified destiny that Pan-African futurists envision, however, is inextricable from network agency born of the Internet, in both liberatory and dystopic ways. As I demonstrate throughout the book, the utopianism evident in the packetization of the neoliberal Internet typically ignores a basic sense of inequality about its deployment in Africa and the underdeveloped world more generally. If the structures of global ICT can be shown to be as enabling as they are limiting and controlling, labeling these practices as Pan-Africanist or decolonial is consistent with the ideals of 20th-century struggles for liberation.

DECOLONIZING THE NETWORK: GHANA AND ITS DIGITAL DIASPORA

The diasporans and digital activists profiled in this book are social actors who have seized the zeitgeist of the information society to produce contemporary social and political movements that resist the narratives of technological abjection plaguing African representations at large. Innovation in this space has been a necessity. Despite the public discourse about the empowering capacity of the Internet over the past 30 years, the network society is hardly seamless or ubiquitously connective. Pervading structural divides and persistent information practices such as "digital redlining" perpetuate unequal exchange between Africa and the rest of the world, not only separating the continent from the techno-modernity of the industrialized North, but also dividing émigré Africans living in the metropole from fuller integration with the lifeworlds of their compatriots in the homeland. Until recently, digital redlining[31]—the deliberate or implicit exclusion of certain regions from participation in Internet economies—has been disabling to users seeking to participate in e-commerce and social media networks in Africa and Latin America.[32] As a case study in the field of STS, this book documents the specific ways Ghanaians are overcoming the asymmetrical structure of the information age to assemble meaningful networks for empowerment and opportunity via novel digital techniques and practices. But the tactics, optimism, and broader political and economic ramifications of such a collectivizing discourse around technology signifies broader implications for Pan-African futurism outside of Ghana. Africans at home and abroad are engaged in similar sociotechnical projects infusing technology practice with a redemptive vision for the continent. Examples include the state-driven industrial capitalism of Rwanda, where assembly factories are making cars and mobile devices aimed at the broader African consumer market; government, nongovernmental organization (NGO), and private finance capital has energized "pan-African" tech hubs in Nairobi, South Africa, and Morocco. Nigeria's ccHub has been a space of innovation for start-ups and activist-developers since the early 2000s, along with Kenya's well-known iHub in Nairobi. Since 2013 iSpace in Accra has facilitated developers and activists across the continent and boasts a classroom/pavilion named after Kwame Nkrumah. Social justice movements have also seized on transcontinental solidarity, emphasizing the insurgent role for tech in Kenya's "Gen-Z protests" (#RejectFinance-Bill2024), which turned bloody when protesters stormed Parliament and police killed more than 20 people. These sentiments find precedence in South Africa's student-led #FeesMustFall movement, which iterated between 2015 and 2016, and the decolonial #RhodesMustFall movement. Nigeria's #EndSARS anti–police brutality movement galvanized international support in 2020, echoing the #BlackLivesMatter movement in the

US, while seizing upon new tactics, such as the strategic use of crypto-currency to fund dissent.

For African activist-developers, these forms of tactical mediation or *calibrations*, as described by critical theorist Ato Quayson, are efforts to advance freedom in societies continuing to endure long effects of European colonialism on Africa's infrastructure and political economy. Africa is the least connected continent in terms of Internet bandwidth and access to the international sea cables that wire this system across the globe.[33] There are fewer data centers and supercomputers in Africa than in any other world region, and many African web services rely on cloud management from Europe, the US, and Canada.[34] The work of Pan-African futurists therefore challenges these persistent economic dependencies and digital divides that further alienate Africa. By focusing on Africa's tech ecology, I hope to reposition the West's digital ecosystem as a particularistic arrangement made global, rather than a universal system: the Internet's capacity for *horizontality* is often embedded in the discourse of the network society, in which "place" has been made inconsequential to data flow.[35] The West's Internet, however, is an ecosystem with its own distinct arrangements of electricity, communications hardware, software applications, data practices, and a corporate and legal regime configured for the most by advanced industrial and consumer economies of the Global North. As described by my research participants, the uneven experience of linking this information system to and from the infrastructure of the underdeveloped world forces users to innovate. These asymmetries have their root in the world system of capitalism and systematic disruptions of colonialism but are experienced in unique ways,[36] as I illustrate in my description of Ghana's cyberculture (see chapter 1). Thus, as innovators and lead users who are configured outside the digital infrastructure of the West, Ghanaians and other Pan-African futurists are engaging in technological innovation to overcome asymmetries in global flows.[37]

These tactics take place amid Africa's inscrutability as a site of technology and modernity. Anthropologist James Ferguson famously described Africa's agency in the contemporary world as veiled under "global shadows."[38] Jenna Burrell described the work of enterprising programmers and entrepreneurial scammers in the cybercafes of Mamprobi in Accra and African Internet users en masse as operating as "invisible users."[39] Like the aspirational narratives embedded in the work of contemporary scholars examining digital media in Africa and the Global South,[40] this book challenges generalizations about Africa's subordinate role in tech innovation, often tied to the epistemological foundations of our notions of technology, as discussed further later. Africans and those from the Global South are often represented as having contributed little to the network society. Instead, notions of Africans' participation in digital culture typically include an undue focus on e-waste recyclers at landfills, email fraudsters, trolls for hire, and the human rights violations in coltan mining.[41] These depictions reinforce the racist notion that Africa exists outside of modernity and fuel narratives of the West's and increasingly China's exceptionalist mission to aid in the external development of the continent through foreign, donor aid. Against these images, the research in this book is in conversation with other Africanist manuscripts intervening in the false historicization of STEM as the province of the West, including Yékú's exploration of *netizenship* in Nigeria, Clark and colleagues' exploration of women's digital spaces on the continent, Donner's examination of the role of mobile phone cultures in development, Burrell's work on Ghanaian cybercafes, Bamba and Miescher's research on industrialization in Cote D'Ivoire and Ghana respectively, Osseo-Asare's description of African technoscience and medicine, Mavhunga's work on African ethnoscience, more broadly Bangura's queries into African mathematics, and Kreamer and colleagues' examination of cultural astronomy on the continent, to name a few.[42]

Ghana has been a key site for transnational flows since the inception of modernity. Formerly known as the Gold Coast, the region was deeply entangled in the Transatlantic Slave Trade. Ghana served as a central site for European empire building, was a base for anti-colonial nationalism and, in the contemporary moment, serves as a site of multiple diasporic formations. For millennia, Ghana, like Africa as whole, has been a dynamic world region, with many of the ethnic groups representing indigeneity today, having emerged from centuries of migrations throughout the continent. The name Ghana itself refers to a "medieval" African kingdom that flourished between the 11th and 13th centuries in the Mandinka region of Mali. The name was chosen in 1957 to harken back to a golden age of independence in Africa.[43] Ghana has historically been a crossroads for diverse African civilizations, from the Sahel region to the kingdoms of the forests in central West Africa and coastal populations traveling in the Gulf of Guinea. When European traders arrived in the 15th century, they encountered emerging coastal cities in Cape Coast and Accra as vibrant ports. The development of the southern coast's sensibility as a "trading zone" was firmly in place, and the incorporation of Europeans into this space further extended the global character of the region for both Europeans and Africans. Ghana's role as a global crossroads was marked and marred by the expansion of the slave-trading economy to the Americas and the Atlantic region in league with European travelers and empire builders.[44] In the ensuing 500 years, as colonialism and political upheaval marked the decline of African sovereignty and increasing dependence upon Europe in the 19th century, the region transformed into a central site for Pan-African encounters in trade, religion, politics, and migration for Africans across the continent.[45] By the 20th century Ghana's emerging independence movement was spurred in part by encounters in Europe, North America, and the Caribbean, as Black elites and religious students traveled abroad for education and training in the metropoles. Students and future leaders studying in the US and Europe began to link their homeland liberation struggles to Black human rights campaigns in the diaspora and to develop a transnational sensibility in relation to the homeland. Thus, Pan-Africanism played a central role in the emergence of the Ghanaian state and others during the 1960s.

Diaspora is a central theme in this book. Despite a strong sense of identification with the nation-state, the notion of *diaspora*—framed by my Ghanaian interlocutors as both an experience of exile and, more broadly, Black unity—has been a central motif in Ghana's national identity. While Ghana's role as a returnee Black homeland dates back to the 19th century, this book picks up the story in the 21st century, examining the ways that ICT, the Internet, and digital diasporas have been central in shaping the discourse of independence, freedom, and solidarity in Ghana, and in Africa more broadly. Digital diasporas and activist-developers have been key actors working to decolonize Ghana and notions of freedom itself from the particularistic arrangement of the Internet amid transformations in technology and globalization that purport to make the world a smaller, more egalitarian "village." The colonial experiences of disruption, or in the language of anthropologist Arjun Appadurai, *disjuncture*, remain a constant for Ghanaians abroad and for those in the homeland negotiating their participation in the world system via an itinerant social and technical infrastructure.[46] The novel techniques examined in the following chapters are analyzed as new forms of technology that attempt to bridge the sociotechnical divides persisting in experiences of Ghanaian national identity between homeland, diaspora, and broader social relations across the globe.

These tools of national cohesion enable participation in the polity of Ghana and are key technologies in the construction of what I describe later in this book as the broader *Afropolitan mediascape*. The copresence of Africans within this worldwide media circuit represents Pan-Africanist

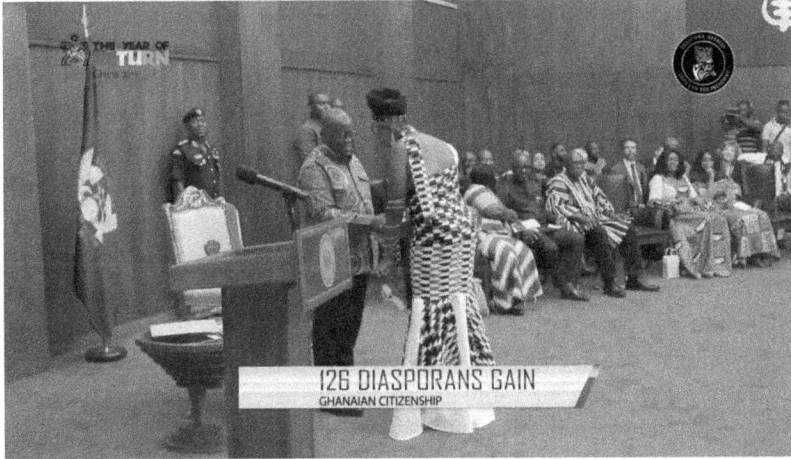

FIGURE 2. President Nana Akuffo-Addo congratulates a woman who received formal citizenship in 2019, the Year of Return, at Jubilee House in Accra. *Source:* Office of Diaspora Affairs, Ghana.

rather than simply nationalist or provincial ties: African films, YouTube-based news channels, online radio and music, and content via Facebook and local media companies such as JoyFM have given way to a global sense of Africa that suffuses all narratives of success in the 21st century. These are evident from regional commercial initiatives such as the AfTFCA, initiated in 2018; decolonial movements; and more politically inclusive civil liberty campaigns such as #Occupy or #FixtheCountry. These tools include the Internet itself, when accessible, and also mobile phones, SMS (short message service), WhatsApp, Twitter/X, Skype, web blogs, and smartphone apps, as well as practices of transgressive computer programming ("hacking") and projects that can be labeled information and communication technology for development (ICT4D).

Due to its history and national character, Ghana and its diasporas are an exemplary case from which to examine the growth and challenges of ICT and development in Africa. Becoming politically independent from Britain in 1957, Ghana led a wave of Pan-African liberation movements for states south of the Sahara in that era, with Nkrumah remarking in that moment, "[O]ur independence is meaningless unless it is linked up with the total liberation of the African continent!"[47] Since then, Ghanaians have sought to embody a cosmopolitan heritage as a regional and global crossroads. Ghanaians have striven to overcome conflicting ethnic loyalties and issues of tribal, colonial, and postcolonial governance, while serving as a host for Black cultural tourism and as a base for Pan-Africanist institutions such as the African Union, ECOWAS, and the AfCFTA secretariat.[48] Following the celebration of its 50-year anniversary of independence from the United Kingdom in 2007, the discovery of off-shore oil reserves that same year, and relatively peaceful transitions between democratic governments since 1992, Ghana has emerged as a standard-bearer of stable African governance in the contemporary era of neoliberalism and globalization.[49] Ghana remains central in the discussion of globalization in the new millennium, with Ghanaian diasporic populations representing significant proportions of African immigrants in global nodes such as New York, Washington, DC, London, Paris, Hamburg, and Amsterdam—key sites in the contemporary Anglophone Black Atlantic.[50]

Ghana's profile has also been heightened in recent years by the emergence of a global African media industry, including web-based news portals; satellite and cable television programming; the global commercialization of Afrobeats, hiplife, and other electronic dance music (EDM) in popular music; and partnerships with the Nollywood film industry of Nigeria, the second-most prolific film industry in the world.[51] The circulation of these culture industries has benefited tremendously from innovations in ICT. For Ghanaians themselves, the concept of diaspora is particularly salient. During the early to mid-20th century, Ghanaians and other African intellectuals seeking status would move out of the continent toward colonial metropoles, becoming so-called *been-tos*.[52] Today, speakers of Akan-Twi, the dominant local language in the country's south, refer to Ghanaians in the diaspora as those residing *aburokyire*, meaning "abroad," in contrast to those living at "home," *efie*. In my research with Ghanaians abroad and those on the Internet, the distinctions between diaspora and homeland are discussed in a similar way. Diaspora is also used as a framework to describe economic development projects, cultural tourism, continental African solidarity, and a unifying global identity for those with kinship ties to Ghana. Until recently, Ghana's Ministry of Tourism was called the Department of Tourism and Diasporan Relations.[53] US census figures estimate that there are more than 200,000 Ghanaian nationals living in the US, with significant numbers in Washington, DC, Atlanta, Houston, the Northeast region, Los Angeles, and Chicago.[54]

Whereas globalization has come to mean the loosening of impediments to freely moving information, capital, and people, this project examines how a collective identity rooted in a nationalist homeland is reassembled by its participants online, in the diaspora, and again in the homeland. In spite of the enduring structural impediments to global telecommunications, commerce, and citizenship between Africa and the West, Ghanaian digital diaspora and media connections are linking its people to more horizontal network flows. As the contemporary idea of "pan-Africanism" is increasingly tied to corporate and neoliberal engagements within the continent, these grassroots digital activists in Ghana and its Black diasporas are reckoning with the notions of liberation and cultural redemption as originally tied to the ideology of Pan-Africanism as expounded by Ghana's first president, Kwame Nkrumah. This book highlights how Ghanaians engaged in these dramatic social changes are negotiating the competing interests of neoliberalism, ICT innovation, and financial globalization, seen as the routes to economic self-sufficiency, against notions of political autonomy and freedom, as idealized in Africa's independence movements. Capitalist bootstrapping narratives underscore most stories of 21st-century success in the postcolonial era in India and China, yet the fervor over Ghana's tech industry and its impact on African innovation and media broadly, I argue, emerges in large part from the Pan-African roots of the nation and its engagement with its Black diasporas. Across five chapters, I use the concept of Pan-African futurism to describe the redemptive vision and on-the-ground projects articulated by tech developers, artists, entrepreneurs, civic activists, diasporans, and returnees, galvanized by connections in digital media and ICT to transform Ghana's and Africa's development.

LIBERATING TECHNE: FROM TOOLS TO TECHNIQUES

In this section I pursue one of this book's major theoretical intervention in STS and new media studies. I argue against notions of technology that move toward what could broadly be described as "gadgets and apps," that is, forms of material culture, such as systems and objects often considered external to the human body, the product of the disciplines and fields of material sciences. I offer instead that the critical site of our investigation of new media and new technology should be the process of *innovation*—creative phenomena that allow tools to produce new opportunities for their users and society at large. What is technology? *Techne*, the etymological root word for the concept *technology*, has historically been defined from Greek as "material practices or the material arts." This contrasts with the much more commonsense notion of technology as objects and tools. I argue, like many in the past decade and earlier,[55] that the transformative element of new technology, therefore, does not lie in tools, but rather in *innovations* that allow the instrumentality of objects to create new possibilities. In this regard, Ghanaian and African social and material innovation constitutes new technologies. That digital technology innovation might emerge from Africa and the Global South, in general, defies for many the sense of both the association of high technology and science with the global "Digital North"—North America, Europe, parts of Asia (China, Japan, Korea, India), and parts of the Middle East—regions where digital connectivity is widespread and ICT is accessible to the majority of the population. African technological workarounds, sociopolitical hacking, digital diasporas, and local innovation in mobile connectivity and now in digital finance, represent modes of new technological production that challenge characterizations of Africa's subordinate role in techno-modernity.

Innovative practices (rather than just tools, systems, and processes) reflect the point at which material transformation is achieved. Often, as historian of innovation Erik von Hippel has argued, this occurs through the introduction of a new technique, such as medical procedures or sporting practices,[56] rather than as wholesale invention. Innovation signals that the crises produced in the social construction of materials have been, in the language of STS theory, "stabilized"—meaning discursive and epistemological arguments about what counts as an appropriate object or technology have been settled.[57]

The word *technology* has emerged from its modernist paradigm of science and development to become a powerful contemporary metaphor for knowledge, agency, and power at both the local and global scales.[58] It has become synonymous with high-tech gadgetry, computer science, and more esoterically, the biological and physical engineering often referred to as the "applied sciences."[59] Indeed, notions of the "network" or "information society" have proliferated since the widespread adoption of computers and globalization in the 1970s.[60]

Theorists in the fields of history of science,[61] STS,[62] and philosophy of science,[63] and foundational Western theorists such as Max Weber,[64] have understood technology to include artifacts such as swords, sewer systems, written scripts, and other processes beyond material goods and machines. These formulations have been helpful for theorizing, so that contemporary discourse and the diffusion of the concept of *technology* to include *tools*, *practices*, and *systems* open the term to a high degree of interpretability. But by equating progress with technological advancement, European and American modernist discourse has justified imperial encounters and historically privileged the tools of Western societies as quintessentially *technological*.[65] Technology, emerging through a machine legacy, was portrayed as the by-product of science, mechanics, and methodological engagement with reason, of which Europe was considered the source.[66] STS theorists have spent the better part of 40 years delineating what constitutes technology in social science research. Scholarly definitions however typically focus on *tools, scientific processes, systems*, and new objects and new practices of adoption.[67] Today, Western chauvinism has largely been removed from most theoretical and philosophical descriptions of technological practice, but the residual bias remains in many forms of discourse including politics, popular culture, and some elements of academic theory. Not only is Africa left out of most accounts of the scholarly history of technology and science, but a public bias endures so that *invention*, whether material or in practice, is seldom seen as occurring systematically on the continent. Nevertheless, the most theoretically relevant approach toward *technē*, or what I take to mean a concern over *practices*, emerges most clearly in STS theory.

From the perspective of Thomas Hughes and other historians of science and technology, the innovation of tools does not represent strictly discrete technical processes or attempts to meet extenuating needs. Rather, technological *practice* reflects the coarticulation of the social and the instrumental via a system-wide process of technological adoption. Hughes's classic treatment of the subject identifies social actors like Thomas Edison as "system builders." In Hughes's historical account, Edison invented the light bulb, devised a system of electrical delivery at the municipal level, and successfully lobbied politicians to adopt this innovation against the predominance of gas and oil-based power and other competing but effective schemes. In the Ghanaian context, one might identify Nii Quaynor as such a figure, having founded the early ISP NCS using satellite connections and moving on years later to become chairman of the government regulatory arm, the National Communications Administration (NCA).[68] Thus, the analysis of technology should not only be observed in the product design and manipulation of tools, but also in tools' sociopolitical context at the individual, community, and structural levels. Edison was successfully able to meet society's and industry's needs with new techniques and maneuver within the sociopolitical milieu to institute these innovations broadly.

Through the work of Bruno Latour, John Law, Michel Callon, and others, actor network theory locates the production of social life via the interaction of humans and nonhumans (including animals and tools). The classic formulation is "technology is society made durable";[69] that is, the interaction between human and nonhuman represents a network, and in this interaction or *assemblage*, instrumental action and meaning are *stabilized* in techniques and artifacts. This stabilization marks techniques and artifacts not only in the instance of a settled scientific controversy, as explained via social constructivism, or the countering of a dialectical force, as formulated in sociotechnical theory. The notion of stabilization can also be applied in the practice of social science, where the researcher can empirically analyze both the actors and the network assembled in instrumentalization of a process or object. In my fieldwork and ultimately in the broader analysis, I encountered Ghanaians, Africans, and their allies engaged in projects bridging the economic and technical gaps between mobile phone users in Ghana and the diaspora; diaspora media companies producing content intended for the homeland; coding classes and start-up workshops administered by volunteer computer engineers in Accra, Kumasi, and elsewhere; and tech entrepreneurs whose business models are either directly or indirectly characterized by a social entrepreneurism (see chapter 5).

During the years of my active fieldwork on the ground in Ghana (2011–2018) organizations such as DevCongress in Accra regularly held mentorship meetings between experienced software developers and youth just starting out in coding. Regina Agyare's tech consultancy firm, Soronko Solutions, founded an all-girls STEM academy that gives school-age girls training in tech and experience working with clients. In the diaspora, organizations such as Black Stars Radio and Adom public access television in Amsterdam, as well as the Facebook channel The Progressive Minds, link the challenges in health, education, and economic development in the homeland to the well-being of those in diaspora. While they are not universal in their complexity, effectiveness, or ethics, I want to illustrate how the innovation among this work's interlocutors is grounded in a Pan-African ethos of liberation geared toward openness and inclusivity in design and intended to create more horizontality, especially for the "bottom billions" in the information economy.[70] As practices of Pan-African futurism, these efforts democratize technology and infuse it with the spirit of Pan-Africanism; they advocate for STEM as a liberating set of material practices for Africans across ethnicities and national borders.

ORGANIZATION

This book begins by describing the deep structural divides that continue to separate Africa from the techno-modernity of the West. I progress through five chapters and a conclusion to examine both the challenges and triumphs that Ghanaian digital actors have achieved to create greater horizontality in the network society during the years of my fieldwork and analysis (2008–2024). In chapter 1, "Asymmetrical Networks: Understanding Ghanaian Cyberculture," I detail the persistent structural divides between Africa and the Digital North. The African digital habitus, then, requires tactical mediation, necessitating that users in Ghana and elsewhere navigate an enduring colonial media ecology in which, as anthropologist Brian Larkin describes, an "infrastructure of breakdowns" is status quo.[71] In chapter 2, "Hacking Development: Ghana's Activist-Developers," I provide a thick description of the work of Ghanaian computer programmers and civil society organizations I describe as *activist-developers,* those social organizers engaged in creating greater political participation in Ghana through skills-based training workshops and business networking. Embracing a developmentalist agenda, Ghanaians and other Africans living in Accra and in other sites in urban Ghana link their mentorship, volunteering, and organizing to a movement for Pan-African renewal, which they characterize as economic as well as political. In chapter 3, "Digital Diaspora," I address the concept of *digital diaspora* as an analytic, providing a social history of the term through classic diasporic and migration literature. Diaspora is discussed as an ethos as well as a technical practice, as a site, and as communities (both online and on ground), through the examination of social media and message boards, and via fieldwork in the Bay Area, Chicago, and Amsterdam interacting with diasporans. I argue that digital diaspora—a concept first explored by Black multimedia artists in the UK—should be considered a technology in and of itself. This includes a constellation of techniques and digital practices, everything from international prepaid phone cards to virtual Ghana cultural festivals held on Zoom during the COVID-19 pandemic.

In chapter 4, "Afropolitan Mediascapes," I analyze the content of Ghana-focused media productions circulated throughout the homeland, diaspora, and more broadly, paired with interviews with content creators in social media and analysis of music and film. I argue that the global flow of these media constitutes their own territoriality, which I have termed the *Afropolitan mediascape*—multivalent circuits or networks that characterize a distinctly Pan-African aesthetic of globalization, one that is embedded in Ghanaian and African approaches to technology innovation more broadly. Chapter 5, "Pan-Africanism or Africapitalism?," examines civic notions of agency and political independence, which are increasingly being tied to economic autonomy, encapsulated in the concept of *Africapitalism*. Here I examine the ways that African mobility and agency tied to tech and globalization encounter what Rita Kiki Edozie describes as forms of "neoliberal autocracy."[72] The conclusion, "The Aesthetics and Technopolitics of a Pan-African Futurism," summarizes ways to think about Pan-African futurism as a cultural and political aesthetic, describing its continued emergence among Ghanaian and African social enterprises in the post-"Wakanda" moment and postpandemic times.

RESEARCH APPROACH

This book can be broadly framed as interpretive social science, drawing on methods of sociocultural anthropology and grounded theory. It is a project of African cultural studies that advocates for an aggressive, interdisciplinary approach so as not to specify African realities for singular interests within one discipline, but rather to speak to holistic truths evident from literary, artistic, and creative industries, alongside field-based research, the determining functions of structural analysis, and the glyphic realities of econometrics.[73] While this work has been produced through both an ethnographic and dialogic frame, with interviews, participation, and media analysis at the core of my research practice, the findings and analysis I am working through rely on theoretical and reflexive frameworks; "thick description"; and analysis within Africana studies, STS, and new media theory. Concepts are framed through language that my interlocutors may or may not share, but wherever possible I have sought to incorporate my research participants in the development and analysis of this work. For instance, while the term diaspora was often used by my interviewees, few had encountered concepts such as Afropolitanism or Africapitalism during the time of active field research. The term *activist-developer* is mostly of my construction, though one may find analogues, such as Wilson and Wong's concept of the "information champion" or 20th-century notions of activist engineers, or "activist entrepreneurs."[74] Ultimately the framing in this book emerges out of my research disciplines and my experiences in the field.

As an Afrodescendant of Black American and Latinx heritage, I simultaneously had experiences as an *insider* for many of my African interlocutors, as my own family experienced episodes of dispersion and dispossession driven by racial exclusion and colonialization. Black American affinity to Ghana in particular has grown out of a rich heritage of exchange and from a shared Pan-African history. Kwame Nkrumah and Nnamdi Azikiwe, two of West Africa's first presidents (Ghana and Nigeria, respectively), attended historically Black colleges in the United States, as had I in my undergraduate years. The notion of diaspora therefore is embedded in this research paradigm, and through my research participants and contacts, this was a concept that was continually worked and renewed. It is more than a term of affinity, as is evident from Ghana's efforts to both instrumentalize its recent diaspora for development projects and draw on heritage tourism as a source of investment dollars. In the discourse in this book and in homeland public discourse, diaspora remains central to notions of Ghana's national identity. My discussions of "The Year of Return" in 2019 are mostly indirect, as I was unable to participate; I look forward to others' ethnographic accounts and critical readings of that key moment in diasporic life.

As with any such research project, building informative and trusting relationships with my research participants has been key to developing an understanding of media use in Ghana's diaspora and in the homeland. Participant observation took place across several field periods starting in 2008, starting with diaspora communities in the US, working online, and then moving on to fieldwork in Ghana and its virtual communities. Some of these consisted of single-day interactions; some were weeks-long engagements; and many of them were disjunctive and multisited encounters with participants both online and on the ground, for more than a decade. My research experiences were episodic at times, visiting Ghana for two- to three-week stints, with trips to Amsterdam and Chicago, which emerged as important and underexamined sites for Ghanaian media production in diaspora. As such this text does not possess the anthropological privilege that a year's fellowship may have afforded. However, I have experienced Ghana much like someone in diaspora, participating peripatetically in the chronoscape of March 6 Independence Day events, national election dramas, and social media and diasporic social gatherings in person, with interviews and documenting online social panics in between. And yet my linguistic deficiency and lack of familial and lived-cultural knowledge of Ghana could never provide a prospective as rich as works such as *Atomic Junction* (Osseo-Asare), *Afropolitan Projects* (Adjepong), and the many works of musicologist J. H. Kwabena Nketia.

While undoubtedly homesick in Ghana during my longer trips, I seldom experienced the deep abiding alienation within Ghana as an African American that have been discussed in other works in which Black diasporans returned to do research. By centering my research in Ghana, I was much less concerned about the questions of African American historical connections as described in texts such as Richard Wright's *Black Power*, Maya Angelou's memoir *All God's Children Need Traveling Shoes*, and Saidiya Hartman's *Lose Your Mother*, or the expert documentation of this by Kevin Gaines's *American Africans in Ghana*.[75] Instead, my encounters could be characterized as largely grounded in Pan-African camaraderie and homecoming, as well as fruitful diasporic exchange.

Utilizing an approach informed by grounded theory and Brock's critical technocultural discourse analysis (CTDA) approach,[76] I have brought several bodies of data into conversation for this book, including

1. interviews and participant observation with Ghana's digital elites: (a) IT developers, social media activists, and tech entrepreneurs living in the homeland, (b) as well as those in the diaspora who are active in online cultural production and transnational media ventures;

2. participant observation with Ghanaian social media users via Twitter/X, WhatsApp, Instagram, and Facebook, participants in the African blogosphere, and other Ghanaian web forums (GhanaWeb and MyJoyOnline, etc.) and on larger digital media platforms like YouTube;

3. digital media content specific to these platforms, including video, podcasts, music, memes, and other online cultural production; and

4. participant observation at key cultural events, during national elections, and at tech conferences and meetups in Ghana, California, Chicago, and Amsterdam.

Analysis has also focused on tools within these media ecologies to examine specific techniques of connection through interviews and participant observation. Content analysis of social media and digital media in general (including streaming content such as films and YouTube channels), in which the discourse around technology, Ghanaian identity, and African identity in general were key points of engagement, has been key to this work. I have sought to document discourse on publicly available digital media, such as open Facebook groups, message boards, and Internet radio netcasts from Amsterdam, where my interlocutors were engaged in public representations. I have included dialogue and discourse from WhatsApp with research contacts with whom I have had sustained relationships. This involved the creation of a large archive of physical media such as magazines and brochures, tech conference programs, festival programs, and photographs. In addition, I developed a much larger archive of digital material, including screen-captured images; video and audio using tools such as Tweetdeck, Hashtagify.me, SnapBird and Evernote; and various web browser plugins. I rely on conventional econometrics for guidance, perspective, and sometimes entry points into the fieldwork, but this project is not database-centric analysis, that is, Big Data.

My on-the-ground research took place in Ghana, at various locales including the Accra metropolitan/capital region, Cape Coast, Kumasi, and Sunyani. In the US, two key sites for my work were Chicago and the San Francisco Bay Area where I engaged in early fieldwork with Jenna Burrell. Chicago's Ghanaian community, while comparatively smaller than diaspora enclaves on the American East Coast and Toronto, Houston, or Atlanta, represents an important and distinctive site. At more than 30,000 people, this Midwest Ghanaian metropolitan community has hosted the annual GhanaFest celebration for more than 30 years. Amsterdam became an important diaspora site in this research, as many of my Ghanaian contacts would refer me to GhanaWeb, an important transnational media company founded and headquartered in the Netherlands. Along with its smaller competitor ModernGhana, also based in Amsterdam, the enterprise and media coverage marked the issues of the Ghanaian Dutch community as an important distinctive community from the other diasporic media producers in this work.[77]

During the 2010 FIFA World Cup in South Africa, the Black Stars, Ghana's men's national football team, gained world notoriety as an African team advancing to the semifinals. This event was significant for me both as a participant observer in diaspora and online. I documented diaspora's engagement via fieldwork in Chicago, while simultaneously networking with Ghana-based fans of football online. Many of my initial contacts on the ground in Accra were active on Twitter/X during those matches, and it is this core group of interlocutors that ultimately exposed me to the activist-developers working with volunteer organizations such as GhanaDecides, GhanaThink, and others. In 2011 I made my first trips to Accra, Tema, Kumasi, and Cape Coast, and in the following year I began making extended stops in Amsterdam to connect with producers at GhanaWeb and the "open access" public media broadcaster SALTO, and with small diaspora media producers. These participants ultimately revealed a repertoire of media tactics aimed at producing innovative connection strategies and social projects between home and the diaspora.

What you will not find in this book are sustained months of discourse analysis of online message boards, which was done expertly by scholar

Victoria Bernal in her work on Eritrea's digital diaspora,[78] or in Anna Everett's early work on Naijanet.[79] Africa's and Ghana's mediascapes are broad and multiplatformed, and like any community's discourse, they cannot be typified by one technology or set of discussions. WhatsApp in particular, which elsewhere I have described as "Africa's platform," has been an important site of transnational exchanges since at least 2014, when it became the most dominant app for users in Africa and the Middle East North Africa (MENA) region before being purchased by Facebook.[80] Though an important site of connection, the closed discursive nature of its (at times 250-member) groups seemed outside of my purview. I have sought instead to document publicly available digital media and discourse, such as open Facebook groups, GhanaWeb message boards, and Internet radio from Amsterdam, where interlocutors were engaged in very public representations. My interviews and sustained relationships with contacts in Ghana, such as participants in GhanaDecides or BarCamps, were more ethnographic in character. The concerns of this project began to push on broader topics as it became clear that *technology* served as a meta-discourse for issues of national identity (beyond the diaspora), the entanglement of notions of development, agency, and modernity.

In all, this project has accumulated more than 200 formal and informal interviews from a diverse pool of digital actors with ties to Ghana between the years of 2008 and 2023. Many of the earlier interviewees include what might be described as experts, what von Hippel would describe as *elite users*,[81] in that they were professionals and activists directly engaged in innovative forms of tech entrepreneurship and social development. I am not arguing that these early adopters are characteristic of all Ghanaian tech use. Rather, like Burrell's work, this research provides insight into yet another underexamined community of global ICT users, but ones engaged in social discourse unique to tech development and social participation in Ghana. Though digital elites in their tactics,[82] many of these actors would not be classified as social elites strictly by their educational background, economic status, and kin relations. Ghana's *digital diaspora* and *activist developers* represent innovative reinterpretations and appropriations of ICT across an asymmetrical "global" IT infrastructure. However, the connection techniques of Ghana's digital majority are centered on Internet access via mobile phones, which was distinct from the cyberculture of the West prior to the mid-2010s.[83] This book highlights a repertoire of ICT practices that produce Ghanaian sociality via new media by working ethnographically across analytical frames for users, tools, communities, and individual actors, as described previously. In also examining home routers, mobile phones, electricity, and Internet access, I attempt to highlight the information ecology coproducing these transnational circuits.[84] I look to document what Kvasny and Robinson, borrowing from Bourdieu, describe as an *information or digital habitus*:[85] not simply the specific new techniques, but the existing and emerging social structures and relationships in which these innovations are embedded.

As a whole, the focus of this book shifts from early questions about Africa's and Ghana's digital infrastructure to concerns about Ghanaian technopolitics in 2023–2024 toward the end of the book. Discussions of contemporary social media use are addressed, including the use of WhatsApp and digital finance projects under the banner of "mobile money," and at the same time, the analysis focuses on Ghanaian Twitter/X and nation-centric content producers and platforms such as GhanaWeb that were active in the 2010s and remain relevant today. The difficulty of writing a manuscript such as this, which addresses contemporary history and political developments and the use of digital media, is that culture and current events can typically outpace a sustained book-length analysis. Rather than chase a presentist approach by analyzing only the most contemporary digital media practices, this text provides insight across a range of practices and social actors that have set the stage for many digital media ventures just emerging now and just assuming mainstream usage (e.g., TikTok, podcasting, blockchain, or AI). However, careful readers of this text will see the cyclical nature in which sociotechnical controversies and political developments from earlier moments of Ghana's (and Africa's) technoculture reemerge, as new digital technologies, platforms, and users adopt and innovate in the mediascape. The issues of postcolonial governance, infrastructure and exclusions, unfortunately remain persistent within Ghana, despite the introduction of both the reformist work of activist-developers and technocratic ventures by the state to increase technology access and entrepreneurship. These ongoing challenges are key to understanding the paradox of technology and development this book explores. This is also why the text itself focuses on the Pan-Africanist sensibility among the social actors, as this remains a persistent moral ethos, rather than only focusing on the impact of one set of technologies.

. . .

In my early years as a scholar of the African diaspora, I was often struck by how heavily rhetorical and discursive academic interventions in the broader diaspora failed to include on-the-ground experiences of Black subjects from the continent.[86] This has certainly been reversed in the years since my research began, with important contributions across disciplines.[87] My interest in Ghana as a site of inquiry stems from West Africa's pivotal role in the creation of broader African diasporic movements and how the region, and Ghana in particular, shapes discourse around social formations across the globe. This work pushes research on diaspora in the African context toward a multisited empirical analysis and attempts to treat Africa and Africans not as a static or essential cultural entities, but as a dynamic and contemporary place and people central (and not peripheral) to questions of modernity and globalization. Key transformations in technology have been central to the African experience since ancient times; they have shaped the development of the industrial era and are now providing key answers to how society will adapt to mobile and pervasive computing technologies, especially in the Global South. This said, the material documented here represents Ghanaian cyberculture in a particular moment prior to 2020, before smartphone adoption began to rise past the 20 percent mark, and just as mobile digital wallets became mainstream in African life. It is my hope that this research will further understanding of how transnational actors and exchanges are shaped by new forms of technological innovation, and how these agents, in turn, shape material practices in general. By highlighting the asymmetries and inequalities in the information society due to geotechnopolitical concerns, this project reiterates the importance of *place* in global networks, as Ghanaian digital actors must continually negotiate new technology against this asymmetrical arrangement.

The confluence of concerns around issues of identity as mediated by ICT are broad at the beginning of the 21st century, with implications for migrant populations and emerging global polities organized around notions of a collective nation-based identity. This research is therefore a case study in globalization, one that reflects a greater discourse around mobility emerging from mid-20th-century postcolonialism and not limited to specific political movements, but to millennial practices of the translocal. By mapping both the practices and cultural concerns embedded in specific networks of global exchange centering Africa, this book provides insight into the ways the continent and its diasporas in the West innovate with digital media. Rather than simply repeating the ways new technology furthers a global digital divide, it is my hope that this research illustrates how ICT provides space for the emergence of diverse knowledge systems linking global resources to local values and to individuals. Rather than imagining something as chimeric as a cyborg from a science-fiction epic, I use the term *cyberculture* here to point to the enmeshing of computer networks with social feedback systems. Today these consist of a range of networking technologies such as mobile phones, email, websites, social media, and mobile phone apps, and more lavishly, an "ambient" Internet of sensor-driven systems, smart homes, public works cameras, and geospatial technologies. Cyberculture explored here, however, accurately describes the entangled feedback of communication systems with the human practices that give them meaning. Since the early 2000s, the principal tool in the production of such a Ghanaian and African cyberculture has been the mobile phone and, despite the pronouncements of cyber-utopianists, this tool has been more successful as a connection technique than Internet accessible via in-home terminals, commercial and public cybercafes, and computer learning centers.

The implications of the research in this book for the fields of Africana and new media studies, in which my work is situated, are broad. Diaspora continues to be a foundational concern in the field, and along with the notion "homeland," this books approaches both as evolving concepts. In the first two decades of the new millennium, these phenomena are increasingly tied to digital practices, as documented by Laguerre, Bernal, Candidatu, and others.[88] These practices change and challenge the relationships of production and consumption amid globalization between the Global South and the Global North. While disadvantages remain entrenched, these flows may reflect new epistemic developments in what Tom Boellstorff describes as our current "Age of Techne."[89] In grounding this book in an exploration of technology practices within the Global South, this book seeks to continue to challenge default notions of innovation as almost exclusively the province of the Digital North. My work with tech developers in Ghana highlights the very contingent conditions with which ICT work in the Global South must contend, in terms of the social, materiality, and systems. This implicates both the practices and sensibilities of ICT development as being continually entangled in the colonialism of the non-European world as well as racist beliefs about technological adoption and progress. In this book I shy away from using concepts such as tinkerism and bricolage to describe the work of Ghana's tech workers. While the term *bricolage* in particular has a more agentive history in the writings of Walter Benjamin, the term in social science has been used to describe the work of untrained mechanics and laborers in Africa and elsewhere. Anthropologist Claude Lévi-Strauss in particular describes this cobbling together of tools as the work of "the savage mind,"[90] in order to distinguish African workmanship from the scientific thinking of Western engineers. In recent years, the notion of bricolage has been reclaimed by hackers and nonelite tech workers to validate the contributions and communities that have emerged from amateur, leisurely, and playful aspects of maker culture and citizen science in the West. I call attention to this discursive distinction both to highlight the ways that scholarly writing should be better informed by research with technological agents in regions and societies whose contributions to innovation have seldom been recognized and to rethink the ways innovation itself has been made provincially Western. *Pan-African Futurism* endeavors to be decolonial, not necessarily in its articulation of an emic approach to ICT based on indigenous philosophy,[91] but rather in its documentation and exploration of innovative practices within African and global techno-modernity.

CHAPTER I

Asymmetrical Networks

Understanding Ghanaian Cyberculture

If the water doesn't work in Ghana, why would the Internet?
—Statement to author by research contact at Hacks Hacker-Ghana
Data Journalism Bootcamp, Accra, 2012

Over the past 30 years, champions of the Internet's democratizing influence have declared that access to global digital information networks "flattens" the world, making information access horizontal. Rather than legacy, analog systems, in which access to information or knowledge is bound up in hierarchical arrangements, in the information society digital connections between nodes in these emerging "interconnecting networks" are theorized to represent more egalitarian relations: a social and technical system composed of co-equal nodes. Both theory and sentiment seemed to bolster the utopian claims in the post-1960s decades that advancements in computing and globalization were leading toward a new age of libertarian freedom through information access.[1] In the 1990s sociologist Manuel Castells labeled this the "Network Society." In 2005 journalist Thomas Friedman naively proclaimed "the world is flat."[2] These predictions have yet to pan out. The world today, connected as it is through digital media, is laden with enduring inequalities in access, capacity for use, and knowledge about the Internet's linking and amplifying affordances by its potential users. Despite the use of naturalistic metaphors about its ubiquity (e.g., "streaming" or "the cloud"), the Internet's infrastructure is largely a physical network that, once mapped, often reveals its deeper divides. Rather than a fisherman's net of symmetrically distributed points, the Internet, constructed primarily through sea-to-land fiber-optic cables and cell phone towers, is marked by drastically pronounced asymmetries between users in the Digital North and those in the Global South, including Ghana and the African diasporas. These asymmetries, typically characterized as a digital divide[3]—a concept that emerged in the US to characterize racial and social inequalities associated with early Internet access—have been the forces shaping the social and technological approaches to digital connection for Africans.

Images of undersea fiber-optic lines produced by the mapping service Telegeography and others since 1999 provide a visceral image of these forms of network asymmetry across the planet. These underwater sea cables, or "tubes," account for more than 90 percent of the world's intercontinental Internet connections, and the vast majority of its high-speed bandwidth wires link regions of the Digital North to each other.[4] Such maps show a dense chord of fiber-optic lines between North America and Europe, numbering more than 17 active cables and segments, while 5 lines run down the entire western coast of the African continent, and only 2 to 3 lines trace the coast from Kenya to South Africa.[5] In 2018 the entire continent of Africa had one-quarter the capacity for Internet connections to the outside world that Europe and North America had to each other and to Asia. These data sources and the UN's International Telecommunications Union (ITU) show that global data traffic to and from Africa via these sea cables lines was recorded at 11.8 terabits per second (Tbps) in 2018, while the largest corridor for international bandwidth transit—between Latin America, the US, and Canada—had more than 43 Tbps.[6] By 2025, despite leaps in African connectivity with more cables, the divide had only grown steeper, with estimates that the North America-to-Europe Internet sea cable bandwidth usage was now five to eight times larger than Africa's international bandwidth connections to the outside world.[7] The sheer number of these tubes demonstrates an innate asymmetry of data to and from the African continent.

The result is that the African Internet, ostensibly servicing more than 1.3 billion people, is slower, more costly, and more highly concentrated in a few regions than anywhere else in the world. While bandwidth and cost of use have continued to improve, these asymmetries particularly had an impact on my interlocutors, especially during the active periods of fieldwork in Ghana between 2011 to 2018, and continue to characterize Internet inequality across the continent, particularly Africa south of the Sahara. Because of these constraints, often obscured by default notions of the *horizontality* of the network society, Ghanaians and other Africans must instead *calibrate* the way they approach their use of these digital networks. This implies not only developing novel workarounds but also negotiating their relationship to the network itself and

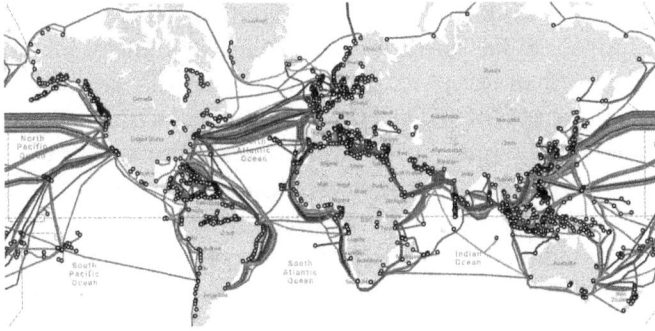

MAP 2. A map of sea-based Internet fiber-optic lines showing the limited connections between African countries and the outside world. The circles indicate cities where "landing stations" are based. Sea cables, or "tubes," facilitate nearly 90 percent of all Internet traffic internationally. *Source: Telegeography*, maps accessible at https://www.submarinecablemap.com/.

engaging in what art and technology theorist Rita Raley describes as *tactical mediation*.[8]

Within Ghana, this involves also selecting and using tools according to the kinds of issues that users on the ground face as they access these networks—including navigating electricity shortages, scarce supply chains for networking tools, comparatively higher costs to access from ISPs, and intermittent coverage. Conversely, in the diaspora, those living abroad must calibrate their usage habits to the tools and modes of connection more commonly used within Ghana, not necessarily those favored in the West. Subscription fees, expensive phone lock-ins, and high international calling charges challenged Ghanaians living in the West during the 2010s; users in the homeland calibrated new practices, quickly adopting and discarding apps and services to maximize costs and accessibility. Ghanaians I observed in the homeland during my fieldwork alternated between ISPs with ease, "topping up" phone credits to access wireless Internet as necessary, while diasporans that I interviewed in the US, often burdened by long-term mobile contracts and locked into specific mobile phone models and expensive payment plans for international calling, constantly calibrated their connection strategies to match users back home. This continues today as Ghanaian digital actors are constructing novel "spaces of flow" via ad hoc, tactical, and asymmetrical practices, in ways other than those imagined by Western users.

In this chapter I use information network theory and STS to analyze the sociotechnical issues that Ghanaians in the homeland and abroad have confronted as they attempt to iterate (refine and tinker with approaches to) social life through unequal Internet connections since the 1990s. Rather than accepting the myth of Internet's democratic horizontality as often promoted by corporate marketers from Silicon Valley to early techno-optimists of the W.E.L.L. and elsewhere,[9] this chapter outlines how asymmetry has been the default experience in Africa—in this case, the imbalance in data connections between continents—and how these conditions continue to impact its cyberculture today. Between homeland and diaspora, rural and urban, north and south in Ghana, *network asymmetry* reterritorializes African digital actors. Digital elites and those living in diaspora often experience an Internet that is "always on," enabled by high-speed broadband, while digital have-nots in the homeland primarily use costly, slow, and "metered" networks. The ITU characterizes this Internet experience as being "marginally online,"[10] and as the following vignette illustrates, such digital asymmetries require a disruptive ethos to overcome the divide.

CALIBRATING DIASPORA

The tactical mediations of a key contact in my California-based research, DJ Mantse, highlights some of the inherent challenges for Ghanaians in diaspora as they seek to bridge these divides. Mantse moved to the US when he was 13 years old. When we first talked in 2009, he was in his thirties and described himself then as an "African in America." Born and raised in Ghana, he had now spent more time in the US than Africa, yet he might also be considered among the first wave of digital migrants,[11] having adopted several specific techniques to stay connected with his home over the years.

Over the last decade or so, Ghanaian tech users like Mantse in diaspora and the homeland have continually embraced new digital tools in often surprising fashion, integrating them into their communication practices in ways directly determined by the asymmetries embedded in Ghana's digital infrastructure. My research contacts and interviewees pointed out how mobile calls, Skype chats, texting, and Twitter/X messages must be strategically deployed in order to establish consistent forms of digital connection that enable social and familial ties. For these diasporic actors, the central issue in connecting from abroad with those in the homeland is calibrating which tools can be reasonably and reliably agreed upon in order to establish transnational discourse.

Beyond its technical usage in engineering, I use the term *calibration* to describe the tactics of producing new meanings within the constraints of social contexts. This involves both adapting new tools for different circumstances and recalibrating what formerly composed settled social conventions. Literary theorist and African studies scholar Ato Quayson uses the concept of calibration to describe attempts to "wrest something from the aesthetic domain [arts] for the analysis and better understanding of the social."[12] While Quayson uses literary arts and texts to read for meaning beyond the embedded notions of the social, we can extend this kind of interpretation to other aesthetic forms, namely for those *texts* composing the "material arts": tools, technologies, and technological practices. While the Internet was by default designed for the West, Ghanaian users calibrate their opportunities and possibilities for both technologies and the social through techniques such as appropriation, or what scholars such as Ron Eglash and Rayvon Fouché describe as "redeployment" by utilizing tools in new contexts.[13] In these ways, Ghanaian digital techniques have recalibrated relationships with homeland tech users and the diaspora, beyond the territory of Ghana.

In the following excerpt from an interview in 2009, Mantse and some of his associates living in the Bay Area describe to me such a process. Some of Mantse's school-age friends had recently moved to the US from Ghana and had just reconnected in Oakland. In our interview, we discussed several forms of media connections used to connect back to Ghana and throughout diaspora. Some experimented with teletypewriters largely intended for use in the deaf communities (TTY), as well as phone calling cards and third-party SMS websites, in the times before smartphones were more accessible on the continent. But these tactical mediations are all deployed to courier messages between users in the diaspora and homeland via desktop transactions, which represent a minority of the ways Africans on the continent access the Internet. This is evident in the portion of the interview where Mantse and his associates describe the redeployment of the VoIP (voice over Internet protocol) technology MagicJack in the early 2000s to connect Ghana and the United States. The MagicJack was marketed as an alternative to cell phone–based communication for those who were late adopters in the West,[14] especially elderly or low-income users. It consisted of an analog telephone jack attached to a USB adapter, which allowed users to simply connect a landline phone cord to a computer. Mantse and his friends described the utility of sending this tool back home to Ghana for users connected to the Internet there. The MagicJack would allow the user to have a US telephone number that could be dialed directly no matter where they were geographically located, even within Ghana. The benefit was the avoidance of costly international phone call fees, though it was not without its issues.

> *Friend 2:* Let me tell you another way, with MagicJack. You know like most people in America here, we leave our Internet on, we just go. I never turn off my Internet. There's a small device called MagicJack. . . . All the MagicJack needs is Internet. So, it doesn't matter where you take it to. I was even trying to bring mine [here to Oakland], in case I want to make long distance calls. Save on my phone bill, save the hotel bills or whatever.
>
> Okay, if I send it to [family] in Ghana, all [they] need to do is plug it into the computer and he's going to have an American number. And plug it in and he can call anywhere in America for free.
>
> *Mantse:* Let me tell you my problem with that, why I didn't rush to that. I didn't rush to do it. . . . You have to have Internet.
>
> *Friend 2:* You give it to the people at the Internet café.
>
> *Mantse:* How many people have Internet café friends? That's why I didn't do it, my manager. . . . He complained about the same thing. But here's what he did. He has Internet that he bought similar to that "BlackJack" thing, that they use over there. But then he transformed his Internet at his home into an Internet café, so that it would make profit.
>
> *Friend 2:* Like my friend, he already had the Internet, so he just plugged the MagicJack in.
>
> *Royston:* You said you used Skype before, which do you prefer?
>
> *Friend 2:* I use the MagicJack, even the Skype, you have to pay, in America you have to pay. MagicJack for $20 dollars, you done. It's free to call anybody, anywhere you want. Except on MagicJack, you cannot call outside of America, you have to call the 50 states.[15]

This conversation demonstrates the diversity of techniques used by this group of Ghanaians seeking to stabilize transnational ties (including those related to entrepreneurship). It also demonstrates an iterative or tactical approach to establishing digital ties, dependent on individual access that must deal with local infrastructure. Even in this process, disjunctive experiences and discourse remain a constant experience, and these do not always find an easy technological solution. Their iterative repertoires draw on a particular set of aesthetic and cultural practices—practices that take shape as affordances of these tools, as well as of the actors involved in these communication assemblages.

The disjunctive information infrastructure between home and abroad illustrates the asymmetry of these global connections and the need for tactical configuration by tech users across Ghana's mediascape. In the quoted exchange, Mantse and his associates illustrate how their cyber-cultural interactions are hardly unidirectional or uniform, even among digital actors in diaspora. The discussion offers moments of surprise, each compatriot sharing tactics that seem at times typical but also novel and sometimes extreme. The choice of tools, as they describe, reflects personal preferences, and also what social role their homeland connections play in their transnational life, a phenomenon that Madianou and Miller in their work with the Filipino diaspora describe as "polymediations."[16] DJ Mantse states elsewhere in the interview that he prefers to connect with musicians via hi5.com, an early contemporary of Facebook. He used this social network for acquaintances and friends who "did not have an email address," highlighting different expectations of digital habitus between home and abroad. With other friends, instant messenger programs such as AIM and Yahoo! IM were more useful. In our later interviews in the 2010s, Mantse stated that WhatsApp was his number one connection tool, which in its initial release allowed users to also interact with SMS messages.[17] Among the dozens of individuals interviewed for this research, I could list similar heterogeneous and hybrid forms of tech adoption and strategizing for social media. If asymmetry is the norm, calibration constantly characterized Ghanaians' approach to Internet access.

MAPPING AFRICA'S NETWORK ASYMMETRY

In order to understand the challenges that Ghanaian Internet users and users across the African continent face, we must first map out the conditions of its connection and exclusions. The African Internet, ostensibly servicing more than one billion people, is slower, more costly, and more highly concentrated in a few localities than any other world region. Bandwidth is commonly low—often less than 2 megabits/sec (2G or 3G and slower)—which makes for slow streaming of media. Coupled with the persistent problems of electrical instability (*dumsor*), the result is that the Internet is not "always on" in Ghana, making difficult many of the interactive and seamless activities of the web, like watching video, video calls, platform gaming, and uploading and downloading content.[18]

Ghana and other countries have made tremendous strides in connectivity since I began this research: In 2010 Ghana's Internet penetration (a ratio of use in comparison with the total population) was as low as 10 percent; by 2015 it was near 25 percent, and in 2024 it stood at

70 percent.[19] And while a lack of high-speed broadband, especially wireless mobile broadband, was largely the story of the 2010s, Africans continue to experience Internet connectivity in fundamentally different ways than the West. Internet penetration as a percentage of the general population is easily twice as much in the US and Europe as in Africa in general. The ITU reported in 2025 that mobile average Internet data traffic (upload/downloads) in the sub-Saharan African region stood at 3 gigabytes (GB) per mobile subscription. This is compared with 11 GB per mobile subscription in the Americas (dominated by the US), and 15 GB per subscriptions in both Europe and the Asia/Pacific regions respectively. Although each world region hosts significant heterogeneity in Internet usage between their countries due to infrastructure and wealth disparities, the global average for mobile Internet data traffic in 2025 was 14 GB per mobile subscriptions, 11 points higher than the African average.[20]

Africa's mobile phone industry has surprised most observers by becoming the most robust regional market in the world since the early 2000s.[21] Even with the addition of more fiber-optic lines and improvements to technology, the majority of African Internet users have always accessed the web through mobile devices, in contradistinction to desktop Internet dominance in the West. Most use a separate mobile hotspot as their wireless modem or tether their computers to a phone's shared Internet connection. Yet since the early 2000s low-bandwidth 2G and mid-level 3G wireless Internet—the Internet speed needed to stream video[22]—has characterized the majority of Africa's download/upload speed. 4G and higher bandwidths finally became used by about 50 percent of the continent's users by 2020.[23] Smartphone and 4G penetration in World Bank–designated "middle-income" and "lower-income" countries such as Ghana, Nigeria, Kenya, Morocco, South Africa, and Rwanda remains at or above 40 percent, and is in the low digits and teens in economically challenged countries such as Mali or Guinea. Apart from university cybercafes and coworking business spaces, Wi-Fi is largely unavailable. Rather, a low-speed, small-screen Internet characterizes how the majority of Africans experience the web. In some markets, competition between ISPs and mobile operators has driven the prices down, but the intermittent and irregular connectivity of those same operators causes users to cycle between providers in an attempt to calibrate their regular Internet and cellular connections. So while many trumpeted the modularity of African device and service contracts in the early years of Africa's cell phone "explosion,"[24] these iterative connection strategies have been a liability for reliable access to the Internet, and that access comes at a cost. Across the continent, especially in the sub-Saharan regions, Internet access costs nearly 40 percent of monthly GDP per capita, with mobile phones and devices costing on average 63 percent of monthly income. The scenario is much more dire for the poorest 20 percent of many countries (the "bottom billions"), where the median cost of an entry-level device is 120 percent of monthly income.[25] In general, Africa's Internet accessibility indicators remain among the most unequal in the world.

Media- and technoscapes connect the world by ever-changing global flows, but gaps, blockages, and dropped connections characterize a disjunctive set of flows between the global haves and have-nots, internally and internationally.[26] I argue that these disjunctures emerge from a much more subtle but pervasive form of imbalance in which the Internet itself is implicated in digital inequality, beyond those imbalances produced by wealth, proximity to stable infrastructures, and the digital habitus of users. Even 20 years into the 21st century, with cloud-based networks and anytime, anywhere computing, Africa is in many ways still dependent upon a sociotechnical system put in place during colonialism. As the maps indicate, data networks linking Ghana and the rest of Africa to the Digital North are significantly less robust than those that connect Europe to the United States and so forth. In the data arena, hyperscale-data server farms (called "cloud centers"), the likes of which are operated by firms such as Facebook, Amazon, and IBM, are few and far between on the continent, with most cloud-computing centers that serve the region being headquartered in Europe.[27]

Reflecting on the history of the 54 nation-states and external polities of Africa, how else could this global inequality in digital networks have been structured? Since the 16th century, European military and commercial forays into the continent have led to successive waves of destabilization, culminating in the 19th century with full-scale political colonization.[28] Between 1884 and 1886 the Berlin Conference initiated the "Scramble for Africa," prompting political annexation along arbitrary geographic lines drawn up in Europe. Africa's political and structural order was irreparably disrupted.[29] Europeans radically reordered racial relations and fostered African economic dependence on Europe, extracting people and resources to fuel empire building for Whites in Europe, while Black wealth was systematically plundered. A colonial infrastructure variously constructed for domination and extraction continued through the postindependence era in the 1950s and 1960s and is arguably still in place. As Ghana and other African nations celebrated political freedom in the mid-20th century, external trade surpluses still evaded most African countries, with precious commodities such as cocoa, gold, bauxite, oil, rubber, copper, diamonds, and timber accounting for the majority of outward foreign exchange. In 1970 Marxist theorist Samir Amin described how this system of unequal exchange had historically and enduringly created conditions for *uneven development* in Africa in particular, as weakened states commanded low commodity prices, which in turn depressed local development, and political alliances prevented the growth of vibrant local business enterprises. As a world system, capitalism has structurally underdeveloped the colonized regions of the world, mostly in the Global South, with infrastructure such as electricity, roads, and water put in service of this extraction economy.

Working through electricity shortages, irregular blackouts, and inexplicably itinerate mobile networks, users in this network are calibrating through what Nigeria film scholar Brian Larkin has described as an "ecology of breakdowns." Larkin contends that such endemic failures and technical difficulties in public infrastructure are not aberrations, but rather the normalized sociotechnical features of everyday life in Africa's postcolonial states.[30] Alongside anticipated expected instabilities, the Internet's issues are compounded by other exclusionary experiences, such as the practice of *digital redlining*. In Africa and other regions of the Global South, practices by ISPs, e-commerce firms, and transnational financial servicers such as Visa, eBay, and Amazon have discriminated, and continue to do so, by neighborhood or geography, unless users' accounts are connected to European or American billing addresses. In Latin America such exclusions have resulted in *zonas rojas*, areas where copper wire thefts and environmentally challenging terrain are proscribed from the grid.[31] Globally, Africans still suffer from public perceptions that *all* online financial transactions initiated from the continent are laden with fraud and scams—such as well-known email scenarios requesting temporary "advance fee" payments, locally referred to as 419 scamming or *sakawa*.[32] In these schemes, individuals purporting to be African elites, political exiles, or would-be lovers request that collaborators share bank account information to temporarily house monetary transfers. It is unclear that Africans are the only source of these scams, but nevertheless the stigma remains. The result has been the persistent peripheralization of Ghanaians and other users from the continent from the mainstream of the global digital economy.[33] Yet in Ghana, digital media have also created spaces of opportunity for innovators and, encouraged by social movements utilizing the Internet's capacity to link people and ideas, such as the Afro-Arab Spring, to seek resistance to this form of exclusion from techno-modernity through Pan-Africanist ideals.

GHANA AS EARLY ADOPTER

Pan-African futurism has emerged in Ghana not simply from the frustration of its network exclusions, but also from its role as an early adopter of the Internet in Africa and in the Global South—in ways that mirror its role in pioneering political independence in the 1950s. This is important to acknowledge before we continue to explore the fundamental issues that have shaped Ghana's digital infrastructure and the actors' tactical mediations within it. This West African metropolitan center has some of the highest Internet capacity on the entire continent, linking to the global Internet via six sea-to-land cables, including ACE, WACS, GLO, MainOne, SAT-3, and as of the year 2023, 2Africa. It possesses several smaller satellite-Internet base stations, and more networks are planned in the near future.[34]

In 1992 the mobile phone company Millicom provided Ghanaians with access to cellular phone service, becoming one of the first companies on the continent to do so. During that same period, Ghana joined a cluster of African nations gaining access to the Internet ahead of the majority of the continent. The former Gold Coast was the first West African country to gain access to the TCP/IP protocol, the coding system for the Internet as we know it, through a mixture of satellite connections and copper-wire sea cables.[35] In 1995 three companies began providing consumer Internet and data services, marking a turning point in the digital communications industry in Ghana. The first two nationwide ISPs, were Westel and Ghana Telecom—the latter was a state-run postal and telegraph service that was formed following state divesture under the World Bank's Structural Adjustment Program. (GT became Vodafone Ghana in 2008, and in 2023 became Telecel.) A third company, NCS, was granted licenses to run satellite connections to the Internet in the mid-1990s and pioneered the use of voice (VoIP).[36] This privately run telecom firm was headed by Nii Quaynor, a future Ghanaian bureaucrat and leader of Internet Corporation for Assigned Names and Numbers (ICANN), who later helped establish the African Network Information Centre (AFRINIC), an NGO that administers Internet names and addresses in the region. The establishment of Internet and data services by these three firms and the increase in mobile phone handsets set the stage for Ghana to be one of the leaders of digital communications in Africa in the mid-1990s. In 2001 the SAT-3 sea cable, owned by a consortium of Western, Indian, and African operators, landed in Ghana, establishing the country's first fiber-optic connection to the Internet, replacing copper-wire transoceanic cables. SAT-3 was Ghana's first high-speed connection to the global Internet backbone.[37]

In other developments during that time, Herman Chinery-Hesse, often called "the Bill Gates of Africa,"[38] founded theSoftTribe, an early West African software firm, in Accra. In 2001 two Europeans, UK-born Mark Davies and Alex Rousselet from France, started Busy Internet in the Ring Road business district in Accra—often called Africa's first Internet co-working hub and tech incubator. By 2002 Accra would earn the distinction of having more than 250 cybercafes, thought at the time to be more than anywhere else in West Africa, though most shops consisted of simple shared computing, print services, and centers for mobile phone "top-up" credits.[39] Also in 2002, Patrick Awuah, a former Microsoft lead engineer, started Ashesi University College, a private tech- and entrepreneurship-focused training school. The Meltwater Institute, a postbaccalaureate training school for start-ups based in Accra, soon followed, with funding and support from founders of the Norwegian/American tech firm that bears the same name. Since then, Accra and Ghana have become a leading education center for ICT in West Africa, attracting students and talent from across the continent and seeing the emergence of dozens of IT training colleges. In 2003 the Fourth Republic of Ghana partnered with the government of India to establish a tech accelerator and NGO called the Kofi Annan Centre for ICT Excellence (KACE), intended to drive development projects. The center would host some of West Africa's first supercomputers, promote local entrepreneurship, and serve as the national administrator for fiber-optic landing stations in Ghana.

As Internet development and rollout progressed in the early 2000s, Ghana, and Accra in particular, continued to lead ICT development in Africa. In 2010, with a population of just 25 million people, Ghana had six national mobile phone service providers. By 2016 more than 20 companies offered voice and data Internet services in Accra alone, providing people with voice calls and Internet access. According to the GSMA,[40] a key mobile phone service operator association, Ghana in 2024 had the highest mobile penetration in West Africa, at 70 percent, and outranks many of its regional peers, even the much larger economy of Nigeria.[41] Thus, it was only slightly surprising to outside observers when Google announced it would headquarter its first African research lab, Google AI, in Accra in 2018. Though the ICT business environment is more diversified and complex in larger countries such as Egypt, South Africa, Kenya, Morocco, Nigeria, and Tunisia, as well as the offshore Internet and finance hub Mauritius,[42] these conditions and a spirit of openness and stability have attracted interest from a cosmopolitan mix of foreign nationals and Blacks in the diaspora in Ghana's tech scene over the last 20 years.

Rather than being simply an exceptional case, Ghana's successes and systemic challenges serve as a typical case for the many issues of Africa's Internet and digital media infrastructure. Despite these early firsts and having an active local tech industry for much of its existence, the Internet is largely unreliable in Ghana. Across Africa, significant gaps in data bandwidth and digital connections create asymmetries between those on the continent and connections to the outside world.

Perhaps most illustrative and crucial to my arguments are the systemic disjunctures between those living in the homeland and Ghanaians living in diaspora. Based on primary and secondary research gathered for this book, I have found that the challenges in Ghana's sociotechnical conditions between 2008 and 2023 included the following:

- Low rates of landline and in-home Internet access.
- Regular and unscheduled electricity outages that cripple digital networks and limit productivity.
- Concentration of Internet and mobile network access in large urban centers and in the southern coastal regions, at the expense of those living in rural areas, and the northern regions, which collectively account for more than 50 percent of the populace.
- Environmentally temperamental and technically itinerate in-country Internet systems.
- Decreasing commercial competition among mobile phone and bandwidth providers and market consolidation since the mid-2010s.
- A lack of cooperation across network providers both locally and internationally.
- Some of the highest costs for Internet access and digital commodities in the world as a percentage of personal income.[43]
- Very little public and free Wi-Fi.
- Limited availability of computing hardware and networking tools such as routers, smartphones, and connection switches.
- Exclusion from the West's robust e-commerce market by firms such as Visa, Amazon, and eBay—digital redlining.[44]

In contrast, Ghanaians in the diasporas are often thought to be favored because of their physical location in the West (North America and

Europe), where Internet penetration rates near 80 and 90 percent, and "broadband" high-speed data access via 4G and 5G networks makes the downloading of video and other data-intense activities more "frictionless." However, those living abroad are not always well served by the sociotechnical arrangements that mark differences between the homeland and the diaspora:

- According to my contacts, Ghanaians in the diaspora are often expected to shoulder the costs of expensive international calling rates from the homeland, regardless of their economic status and income.
- Expensive mobile devices and data subscriptions tie diasporans to long-term contracts with a generally limited number of national mobile service operators (three in the United States), whose market consolidation has resulted in stabilized high data and device fees, mostly through expensive service contracts—though connectivity remains more stable.
- These same mobile phone and Internet networks, rather than linking to the global Internet seamlessly, at times charge high prices for international calling, at rates up to five times the amount that it costs for someone living in the homeland to connect to those living abroad.
- Prior to the late 2010s, monetary transfers to those living at home were often routed through just a handful of servicers, who charged as much as 20 percent per financial transaction, such as Western Union. While the phenomenon of digital wallets or "mobile money" promises to alleviate the friction in sending home such remittances, the reliable systems remain pricey, time consuming, and administratively tedious.

While the extension of the Internet has made African and Ghanaian media potentially global—with African music, film, and religious programming accessible to diasporans—it is important to note how the inherent digital disjunctures described previously continue to impact Ghanaians and other Africans in diaspora. Ultimately, Ghanaians living abroad are required to participate in a different, if at times less robust, Internet, as they employ connection strategies from the West to Africa. Ghanaians in diaspora who are receiving and making mobile phone calls or social media connections not only are on "Africa time," as my contacts often described—receiving calls and requests to connect at all hours of the day and night—but they must use the platforms and tools that are easily accessible and affordable to those living in the homeland. It is often the digital habitus of the homeland user that determines the Internet praxis for those in diaspora. It is often a peripheral Internet to the one they experience living in the West and at times requires additional cost to their Internet. For those who navigate these connections across global divides, the optimistic promises about the potential of the Internet are not well borne out and are evidence of Africa's continued dependence on the West and its failings as a space of techno-modernity.

THE TROPICAL INTERNET

Similar to the way DJ Mantse in the United States and his associates in the homeland redeployed existing tools to negotiate connectivity with each other, tech developers in the Ghanaian homeland also calibrate their practices to account for infrastructure asymmetry between Africa and the Global North. During my visits between 2011 and 2018, tech users in Accra and Kumasi urban region had access to four additional wireless, high-speed mobile broadband operators, at times giving these bustling cityscapes Internet experiences on a par with parts of the metropolitan West. But outside the city centers and away from the coastal region—where nearly 50 percent of the population resides—Internet and wireless connectivity remains spotty. As Herman Chinery-Hesse would describe to me in 2013, the problem is that the Internet and other large global systems for commerce and communications are not built for Africa's "tropical environment."[45] Chinery-Hesse trained in software development in the mid-1980 and 1990s working for software firms in the US and Europe. He told me that while the Internet's connective potential had inspired his love of coding, reading *The Autobiography of Malcolm X* was just as much of an awakening for him at that time. He returned to Ghana in the 1990s, having reached what he felt was a glass ceiling in corporate coding culture: "In the United States, I was made to feel like a second-class citizen. The American race thing is very troublesome for Africans. Most folks just find it easier to leave."[46]

Chinery-Hesse said that since 2003, theSoftTribe has utilized what it calls a "tropically tolerant" approach toward implementing software and technology in Ghana and its other markets in West Africa. In this approach, as he has outlined in numerous media interviews and online discussions, ICT in the African region is unique in that it has to account for the persistence of infrastructural woes, including frequent power outages and accompanying surges, dust during Harmattan season, and most of all patchy or altogether interrupted experiences on the Internet.

In Ghana, the "tropically tolerant" approach means taking into account many additional specific infrastructural problems, including failing roads; lack of consistent network access for clients and their own business; inconsistent supplies of technical equipment; and the oft-bemoaned experience of rolling blackouts, scheduled and unplanned. Employing the techniques used throughout the early 2000s and 2010s especially, theSoftTribe and other firms regularly "cached" their vital systems, relying less on distributed servers or wireless Internet and more on programs that are designed to save locally to hard disks. Data was queued for transfer at times when the Internet was not heavily trafficked. Programmers wrote to hard drives before distributing or sharing data. Internet sites, especially those containing important technical information and specifications, were saved as local files, so that they could still be viewed in the event that data servers and the Internet itself were not accessible. Thus, the digital habitus of remotely stored data systems, cloud servers, and virtual machines was less typical in an environment where Internet was as precarious a utility as water in this part of the underdeveloped world. Even today, local work-arounds and physical environments continue to be as much a detriment to the Internet experience as lack of access to the tools and technical knowledge about how to make them work.

This was illustrated by my experience as a researcher on the ground. The utility provider the Electric Company of Ghana mandated regular "load shedding" exercises and unscheduled "lights off" periods. The power outages also made cellular broadband service just as intermittent, and the Internet was inaccessible even at commercial facilities such as cybercafes and print shops at these times. Equipment issues often dogged access: at facilities I surveyed in Accra, Cape Coast, Kumasi, and Sunyani, desktop computers were typically 5 to 10 years old, running operating systems that were long out of service from major firms such as Microsoft, Dell, and Intel. A robust trade in secondhand computer devices from the West was obvious, with barcode stickers from universities and public libraries still visible on some units at commercial Internet centers and device sellers. Surveying devices in cybercafes and other locales, I often found storage space on these terminals was limited and processor speeds sluggish, requiring that programs and interfaces use the least amount of computing resources possible. These constraints have downstream effects on the way users connect to networks and, by extension, communicate with users inside and outside the homeland.

As Chinery-Hesse describes it, in order to be tropically tolerant, most software is designed to be backward compatible for a range of devices. For ICT users and managers, literacy in coding and technical training, even English at times, is uneven: tropically tolerant software is designed to accommodate system administrators who may have only rudimentary knowledge of networking, much less of programming. theSoftTribe has described needing to "beta" its releases carefully to avoid both wide system failures and the potential for others to pirate its software.[47] *Tropical* in the late 1990s and first decade of the 2000s thus served at that time, and to a large measure now, as a metonym for the travails of a postcolonial nation struggling with the challenges of ICT development.

EVERYDAY #DUMSOR

Endemic crises in energy are not normally identified with Internet woes in the Digital North. Though generators are a common sight in many cities in the Global South, Ghana's hydroelectric dam and utility system had been the pride of West Africa from the 1950s to recent times. As historian Stephan Meischer has documented, the Akosombo Dam Generator (ADG), an achievement of Kwame Nkrumah's First Republic government, has struggled in the last two decades, becoming a sign of Ghana's postcolonial malaise and tenuous infrastructure.[48] Ongoing failures of the national electric grid started in the 1990s and continued at key points in the early 2000s, due in large part to the dwindling water levels impacting its capacity. The dam suffered from deteriorating equipment. Poor transmission technologies and damaged electric lines from collapsed roads have slowed access. A growing population in urban service areas such as the capital, Accra, has also increased stress on the electric grid. As a result of the many problems at the dam, the Electric Company of Ghana's (ECG's) scheduled and unscheduled blackouts have become more routine in the last 15 years. Since 2000, water-driven electrical generation from ADG began to decrease as a result of global warming and increased demand from growing southern urban centers.[49] As Mieshcer stated about energy crises in the 1990s, "As the nation watched the water level of Volta Lake drop, Akosombo's role as the engine of Ghana's modernization came into question."[50]

The local word *dumsor* (pronounced /doomn suh/), meaning "lights off, lights on," describes the increasing daily phenomenon of electricity failures. "Lights off" previously peaked in 2008, when dumsor was used sparingly, according to research participants I interviewed for this book. But the term returned with greater regularity when an oil pipeline in the Gulf of Guinea that was supplying fuel to Ghanaian auxiliary generators was ruptured in the summer of 2012. To deal with dwindling hydroelectric power, gasoline was imported to power the electrical grid at thermal generation plants. (Ultimately, then President John Dramani Mahama would import two floating generators on barges to supplement power generation in the lead-up to the 2016 election. He lost that year to Nana Akufo-Addo.) The rupture of the pipeline made the frequent, though irregular dumsor in the urban core a daily reality. Gas-driven generators began to proliferate, as did confrontations over the inability to acquire electricity, including theft and related crime. By 2012 hashtags criticizing dumsor in Ghana began to center around ECG's rising prices and inability to provide lasting and basic service. The use of #dumsor soon became a *transmedia* sign, migrating from common speak to social media, where its hashtag became a way of signaling an ecology of breakdowns and the failure of the state. The despondence the term also carries within it has been potent in forcing the political establishment to deal with infrastructure problems that betray Ghana's narrative of African exceptionalism, particularly with regard to peace, development, and democratic social order. In 2015 dumsor outages and torrential seasonal rains contributed to the explosion of a GOIL gas station in downtown Accra that summer, killing more than 100 people and wreaking havoc near the Nkrumah junction. Protests and calls for accountability dogged the government in those years, and Mahama's administration tried to embrace the hashtag, before ultimately being voted out of office.[51]

In November 2012 the growing use of #dumsor in social media was not yet accompanied by the street protests on the part of everyday Ghanaians or civic activists that ensued in the months after the explosion. But the frustrations of intersecting breakdowns in civic and digital structures was obvious among Accra's bourgeoning stock of high-tech programmers as power outages shut down their business activities and caused enormous frustration. On a humid afternoon under gray skies, I sat down with Anne Amuzu, CEO and cofounder of Nandimobile; we discussed the state of Ghana's developer scene. In 2011 Nandimobile had surprised the global tech community, becoming a prize winner at the global start-up event LAUNCH festival in San Francisco, a first for Africa and a Ghana-based start-up. Amuzu, normally busy running the customer relations software firm, usually would not have time to discuss tech in Ghana with curious foreign researchers such as myself. But dumsor had struck the Meltwater Institute,[52] the tech hub and training school her firm was working out of, and electricity was down in this usually upscale and "protected" neighborhood, called American House. I tried apologizing for interrupting her day, to which she quipped, "No worry there's no light, so I may as well sit here." Power had been out for hours, and an engineer was called to service the two backup generators, which were also down. No business or training sessions were happening for the more than 20 small firms working in this tech incubator. Although Amuzu was a spirited young CEO known for her optimism, the "lights off" nevertheless left her feeling dour. "It shouldn't take long but it does," she griped at the time.

As we sat on the balcony to escape the heat, our conversation about Accra's tech developers turned to the question of transnational connections and the improvisational methods of overcoming these and other structural impediments that plague Internet users from Africa. Amuzu, like many I spoke with, enjoyed the relative stability of Skype video chat at that time but seldom used its text-messaging feature to connect with others both locally and abroad. "The problem is, the SMS feature, you have to pay for it. You need a credit card or debit card. Most people don't have access to that, so it's more difficult to pay for the SMS," she explained. In just a few years, digital wallets by telecoms such as MTN would be widely adopted. "Some people have also blocked us because of the fraud thing, from paying from Ghana with your credit card."

Users in Ghana in 2012 were severely restricted from downloading apps as a result of digital redlining. For legitimate business users and the public, work-arounds were essential for participating in some of the most basic e-commerce practices, as Amuzu explained. "Most people either have their phone or iPad jailbroken, or they have someone out there [in North America or Europe] who has an account and they're using it with the person. They get their details. You just need their user ID and an address. You can even synch it to your device. So some people go around the system. When I was trying to get an Apple ID, I just used some random US address. Even then, you can't even use your credit card details. You can just get the free apps. There's no straightforward way."[53]

Today there is a simple solution to this issue: mobile money and digital forms of currency, in which mobile service providers facilitate person-to-person financial exchanges via mobile devices without the aid of banks. In another bright spot, Africa has been the world's leading adopter of these and other person-to-person e-commerce features since 2016. In 2022 the GSMA estimated that sub-Saharan Africa accounted for over $700 billion in digital wallet transactions, compared to $68 million worldwide in 2012.[54] While developments of cryptocurrencies are outside the scope of this book, digital wallet services such as East Africa's M-PESA had been widely adopted in Kenya at the time of my fieldwork in Ghana. But in 2012 mobile money services were hardly robust in Ghana, with some users even quipping to me that they were more likely sources of fraud. During the early 2010s and even today, in order to bridge existing digital financial divides, diasporans and homeland Internet users calibrated, as Amuzu and others described, utilizing limited networks in order to stabilize their global connections. This practice is somewhat similar to what Mirca Madianou and Daniel Miller described among Philippine transnational families in their ethnographic work as *polymediations*. In their book, the researchers describe how actors are "constantly changing media and the need for each relationship to create a configuration of usage generally employing several different media" to connect between parents, spouses, and children separated by labor migration.[55] These contingent and at times ad hoc methods demonstrate both their commonality and ephemerality throughout the Global South. In my early visits to Ghana between 2011 and 2013, expert users often relied on tethering their laptops and even home computers via a USB cable or wireless Bluetooth in order to sustain a useful Internet connection through a smartphone's cellular Internet access. By the late 2010s new wireless broadband firms such as Surfline and the Busy Internet mobile company began to market "wireless hotspot" devices for high-speed Internet at 4G/LTE (3 to 6Mbit/s), providing yet another layer and device in a professional web developer's or networking engineer's bag of Internet connection tools in Ghana.

Even among Ghana's *digital elites*—expert users, Internet entrepreneurs, and social elites with disposable capital—many described having to use a constellation of connection strategies that included the tethering technique outlined here to deal with the Internet's temperamentality. Still, my contacts complained often that service providers' signals would degrade as the day progressed, with some experiencing weeks of unexplained lags in network speed. These small and medium-sized businesses would often double and triple up on their ISP connections. Redundant ISP subscriptions, however, would be necessary, as users switched between services depending on the time of day and sometimes based on their location throughout Accra. Henry Addo, an early software developer for the social enterprise Ushahidi,[56] a service used to map crises throughout the African continent and world, told me in an interview in 2011 that when Internet access became problematic for days on end, he would move from Accra to other cities in Ghana. He discussed even relocating to his colleagues' offices in Kenya to ensure reliable access.

The experiences of another computer programmer and entrepreneur, Emmanuel ("Emma"), illustrate the tenuousness of Internet access in the homeland, even for elite users. When I began interviewing Emma in 2011, he had recently relocated from Washington, DC, to start an IT consultancy in Accra. Emma lived in and operated out of Madina, an area north of Accra's city center that seemed to him to be a "developing neighborhood" with increasing access to services. But landline telephony in Madina was, in his words, difficult; Internet reception was spotty; and problems with other utilities, such as running water and electricity, made it difficult to maintain operations. Following a particularly torrential rainy season, we sat together looking at the roads outside his office/apartment, pockmarked with impassable trenches and sinkholes that were not there just a few weeks earlier. Despite having an address in the capital region, Emma's means of connecting to ICT was limited by the specificity of his locale in the city. To do his work as a network engineer and software developer, he needed to be constantly attentive to the instability of environmental factors in the existing Internet infrastructure. To compensate, he purchased four subscriptions from competing ISPs in order to ensure constant online access for himself and his clients. This maneuvering included one ISP that used a "line-of-site" radio transmission for access to a hardline fiber-optic tower, as opposed to the telcos commercial cellular or landline connections. "As a Ghanaian, you have to make a way for phoning and ICT. You have to have multiple Internet dongles. You have to be a land and business [dish] subscriber for Internet services. Redundancies!" he said, shaking his head in frustration.[57] After a few months, Emma wound up moving his office closer to the central Osu business district.

RETERRITORIALIZING THE AFRICAN INTERNET

Networks mark both points of connection and geographies of disjuncture. The physical networks that enable the Internet connections between Ghanaians in the homeland and the outside world, including its worldwide diaspora of more than 2 million people, are both instruments of techno-modernity and boundary-marking devices. In media and technology studies, deterritorialization has been a key if often peripheral idea used to understand the Internet. A post-structuralist notion advanced by Gilles Deleuze and Félix Guattari as a critique of late capitalism and psychoanalysis in the early 2000s, it was employed by key figures such as Appadurai, Garcia Canclini, and Pierre Lévy to describe the delocalization of media structures, among other phenomena of globalization. In political-economic terms, deterritorialization can describe the process by which value is abstracted from its context and inserted into new positions (reterritorialization), to better effect capitalist extraction or colonial notions of use-value.[58] For many scholars of globalization, the deterritorialization of media organizations, distribution channels, and content production has been invigorated by ICT—allowing foreign and diaspora audiences to share in a public discursive and media space within otherwise bounded national and local contexts. In this process of globalization, geography-based cultural flows become "untethered" from ties to land and are transformed into what Manuel Castells calls the "space of spaces."[59] Networks that have emerged among Ghana and its diaspora provide a powerful example of this kind of cultural flow, reterritorializing Ghanaian media as more-than-national flows via global Pan-African political, cultural, and digital circuits. Pierre Lévy, in the influential text *Cyberculture*, states that the great advantage of having systems of email and satellite-connected media is the resulting deterritorialization of knowledge from geographic centers (in the West) and in hegemonic centers globally. He states, "[T]here can be no virtual community without interconnectivity, no large-scale collective intelligence without the virtualization or deterritorialization of communities in cyberspace."[60] Whereas the metaphor of the network society assumes as its ideal what Castells describes as a *horizontal* arrangement of group relations, as the case of Ghana and its diasporas illustrate, disjunctive relationships of coloniality remain embedded in homeland-diasporic digital circuits, at times enhancing deterritorialization due to Internet asymmetry in Africa. These network conditions and tactical media practices configure relations in systems that are disjunctive and not simply "flattened" for users and agents. The distance is real for Ghanaian Twitter/X users checking their feeds daily from Amsterdam or Chicago, or in Accra, where developers and their clients wait for the electricity to come back on.

Networks are models for human organization; that is, they are abstractions of the messiness of real life. While so many of the core ideas about flow, flexibility, speed, and response in the network society have been borne out, Castells's early optimism about the democratic ascendance of ICT practices should be tempered with real experiences of network inequality on the ground.[61] The oft-cited origin of this theoretical model for a *distributed network* is a diagram developed by RAND corporation scientist Paul Baran.[62] Developed in the 1950s as a hypothetical arrangement of command-and-control systems for the US military in the event of nuclear war, Baran's lattice-structured distributed network would designate each computer-enabled military outpost as a point in a

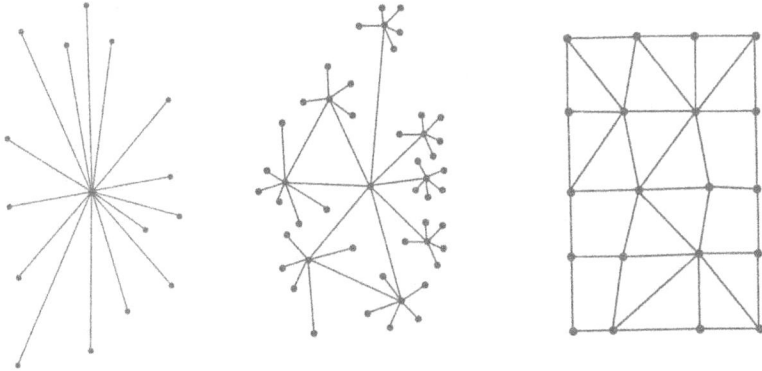

FIGURE 3. A rendering of Paul Baran's notion of (left to right) centralized, decentralized, and distributed networks. The Internet is theoretically imagined to be a "distributed" network (right) of independent nodes. Image by author.

system where nodes could be interoperable. Each station would have the ability to store information on its own and perhaps assume command of the network. The key affordances of this decentralized, grid-like structure would be flexibility and independence, but also collective action. This was the dream of ARPANET,[63] a digital network funded by the US military during the 1960s, and the Internet as we know it was born from these dreams.[64] Yet it is the reliance on these mental models of evenly distributed, symmetrical networks that prevents us from more accurately conceiving of real relationships across the network society.

The essential components of a digital network include two points and the link between them (sometimes called "edges"). There are other nodes of distinction (routers and relays, etc.), but these two elements constitute the theoretical imaginary of network design. Castells reaffirmed an underlying belief in the coequalness of points in the Internet as a distributed network: "Within a given network, flows have no distance, or the same distance, between nodes."[65] The symmetrical flow of information between points in the network, or "horizontality" as he states, is the defining affordance of this nonhierarchical configuration. However, I argue that all networks, regardless of their ability to be distributed throughout a social or technical system, have uneven degrees of power and agency. They are subjective models of relations and also come into being through subjective acts and characterizations. As a globalizing process, these networks may allow communities to exist outside of territory, to be multisited. In Ghana and the diaspora, borders and places are what dictate the experience of connection on either side of these asymmetrical networks. Territory matters. Networks matter. Place matters.

For comparison, in the United States, with its deeply overlapping cable and telecom infrastructures and multiple redundancies from competing networks, there is a sense that the Internet is "always on." Key practices of the Internet in the West depend on this always-on availability, as well as high-speed bandwidth—4G or above, and 100Mbps download speeds.[66] Consider services such as on-demand video (YouTube, Netflix); video chat (Zoom, Google Meet); social media (Instagram, Twitter/X, TikTok); streaming audio (Spotify, online radio); and cloud-based storage, access and information retrieval from servers such as Amazon Web Services. As we contrast the West's Internet culture, long tied to desktops and high-speed landline bandwidth, with the descriptions of Africa's Internet by the entrepreneurs and everyday users throughout the chapter and in this book, it becomes clear how infrastructure at each node in Ghana's global network remains asymmetrical to the Digital North. This is not simply because of existing political-economic relations via the World Bank or low liquid capital among Ghanaians, but due to the network society's hegemonic infrastructure and Africa's comparatively less robust network capacity, which configure the user via regimes of connection. An often-repeated observation by my research participants was that it's easier for people in Ghana to call their counterparts living in the United States than the reverse; US carriers charge significantly higher prices for "international" calls to Ghana than Ghana's telecom networks (such as Vodafone or MTN) charge for these same outgoing calls to the US and elsewhere. But even in this case, factors such as *time, duration of calls*, and *call schedules* are determined by the praxis of homeland users. The various infrastructural pressures and work-arounds that Ghanaians in the homeland experience require those in the diaspora to conform their digital practices to a timeframe and mode of connection that is more favorable and cost effective for those residing in the homeland. In effect, Ghanaians living abroad are on "Ghana time." This space of flows, even within sites such as London, Toronto, or New York, abide by the network logic of Ghana, often extending peripherality and the realities of "breakdown" to the metropole. In turn, Ghanaian diasporans connecting back to home rely on Western infrastructures, which for many years privileged connection with other Ghanaians in diaspora in the West. This is evident from web traffic and postings that I have observed on GhanaWeb and social media such as Twitter/X and Facebook: Online, the diaspora was often more connected to itself on the Internet than it was to the homeland (more in chapter 3).

Conversely, the affordances of high-speed Internet in the West are not always liberating for its diaspora users. Diasporans using mobile phones from the United States, for instance, are often locked into expensive monthly payment plans from one of only three top national cell phone providers (Verizon, Sprint, AT&T). In contrast to the highly consolidated market for nationwide providers in the US, a competitive local telecom industry thrives in Ghana. Price wars often drive users and traffic back and forth between the servicers. In the early period of my work, Ghana boasted six national providers, a number that has dwindled to four in recent years due to corporate consolidations. Telecommunications companies in Ghana provide both home and mobile Internet connectivity, with mobile Internet far and away the most-used service. Because of this method of basic Internet delivery, the four major national firms (Vodafone, MTN, AirtelTigo, GLO) can easily be considered the public face of the Internet,[67] in a way that is not readily analogous to the scenario of access in the United States for firms such as Verizon and T-Mobile. Despite there being choices for consumers looking for the best network connectivity, the overall asymmetry of these communicative systems, from urban to rural, from home to abroad, and back, ultimately deterritorializes and then reterritorializes the Ghanaian and African Internet user. The mostly foreign-owned sea cable networks make Africa's Internet largely dependent on Digital North technology and its forms of access. Accessing the Internet from Ghana is expensive and a sometimes cumbersome endeavor, as several quantitative studies bear out.[68] Urbanites and elites are typically privileged users. The rural populace, nearly 50 percent of Ghana as in most African countries, are "marginally online." The novelty of iterative techniques of connection described by network engineers such as Emma may be a source of inspiration for those accustomed to an always-on Internet, but the disjunctures and procedural redundancies required to make new ties and sustain existing ones represent the ongoing challenge for innovators within the African mediascape.

Distributed networks, as the Internet is often characterized, are idealized today for their ability to connect and as a model for hyperdemocratic forms of organization. The presumed "flatness" of these networks can be likened to basic notions of mathematical symmetry or equality. According to anthropologist of technology Bryan Pfaffenberger, the *principle of symmetry* should be invoked during an evaluation of whether an object or process ("a stabilized technique") accomplishes a relevant function.[69] He argues that technologies should be evaluated based on their capacity to fulfill tasks or functions, regardless of whether or not they have been widely adopted. Failing to do so privileges only techniques that have "taken off," because of what can be described as nonobjective values: He states, "Of apparently successful systems, *we can say only* that the system-builders have apparently succeeded in bringing to life one out of a range of possible systems that could achieve its goal. . . . Social choice, tactics, alternative techniques, and the social redefinition of needs and aspirations all play a role in the rise of sociotechnical systems" (emphasis added).[70]

In fairness, the Internet in the West is, in reality, less distributed than Baran's model of a chain-link fence or a spider's web; ISPs like Comcast and Verizon own key elements of the US network backbone and the data distribution centers to which individual users of the Internet connect. And as open as it has been, the US has experienced internal parochialism with regard to its networks: "net neutrality" has been the policy goal for advocates since the inception of the World Wide Web, with groups such as the Electronic Frontier Foundation insisting that ISPs have no legal right to restrict or limit access to, the speed of, and the content on their privately owned fiber-optic networks for their own commercial advantage. These rules were ensconced in the mid-2000s until US President Donald Trump's first administration dismantled such regulations.[71] Such openness was one of the key forms of efficiency in the US network system; it allowed for less active routers and data lines to connect to distant nodes easily, rather than forcing communication paths between nodes to move sequentially along any one given network. This arrangement, with commercial administrators and ISPs controlling access to the core networks, is very different from the image of every device linked directly to a broadly democratic system or something as democratic as an "information superhighway." Access is largely determined by agencies such as the Federal Communications Commission in the US, Ghana's NCA, and ICANN globally, regimes which can, in effect, strengthen borders in the digital arena. With technopolitics at the core, it is easy to see how the principles of embedded inequality in other systems of exchange, specifically the *core-periphery* divide at the heart of world-systems analysis,[72] provide yet another way to understand unequal relations between the network privileged in the West and the network poor in Global South regions such as Africa.

Of the UN's "top 10" Internet economies in 2021, the United States and China were the clear leaders, followed by countries in Europe and

East Asia; African countries dominated the "least-developed countries" list, with the lowest Internet and mobile broadband penetration rates in the world.[73] Structural inequalities are at the heart of the Internet's asymmetries, in which lower-income countries in much of the Global South operate outside of the mainstream sociotechnical experience of the Internet as it is experienced in the West and advanced industrial economies such as China. These imbalances favor connectivity for those operating from this Digital North, linked, as it is, more to itself than it is to the rest of world, especially through high-speed services.

Having given presentations on network access in Africa at conferences for more than 15 years, I am often caught off guard by the haunted reaction that maps of the Internet invoke in my audiences—especially among Black Afrodescendants. For many unfamiliar with graphics showing the world's continents linked through assemblages of fiber-optic lines and undersea cables, such images do not simply trace a line between points of data, but also invoke notions of the "Triangular Trade" that once ferried enslaved Africans between the continent, Europe, and the Americas and (seldom) back. In exchange for crafts and commodities or raw materials and precious metals, more than 12 million Africans were sold into the world's emerging mercantile and capitalist flows, largely directed by European and some African polities between the 15th and 19th centuries, largely using these same maps. This form of globalization and modernity linked Africa to the world in a profoundly new and deterritorializing arrangement, initiating what Walter Rodney and Samir Amin have described as an epoch of Africa's "underdevelopment."[74]

In the 21st century, these transoceanic routes have given way to a new global system, in which extraction of other resources from Africa fuels the "information economy." Today, minerals such as copper, columbite-tantalite (coltan),[75] and lithium are mined from the continent in both legitimate and illicit capacities, to be used in Asia to make mobile devices and fiber. The administration of these processes is owned by and overseen by the Global North, ultimately servicing the planet's consumer electronic industry. In this system, with a few exceptions, Africa is a consumer, not a producer of technology, and so as in times of slavery and colonization, the continent remains a site of resource extraction and dependency. The specter of this history is central to the passion for autonomy and self-reliance that Pan-African futurists pursue. In the next two chapters I examine how these forms of calibrations and outright system hacks disrupt the default network asymmetries, through the techniques and social advocacy of Ghana's activist software developers in the homeland and in conjunction with its digital diasporas. Such a Pan-African futurism utilizes novel digital techniques, mobile apps, social entrepreneurship, volunteerism, and extramural coding literacy as tactical means to forge new flows within this variegated Internet landscape. This discourse and practice draws less upon the desire for Ghanaian nationalism aided by technology and more on the image of global African and Black unity central to the vision of Nkrumah's Pan-Africanism of the 20th century.

CHAPTER 2

Hacking Development

Ghana's Activist-Developers

In Africa, we know how to hack things.

—Omoju Miller, senior adviser to CEO
of GitHub and machine learning expert
with roots in Nigeria, stated during online
participation at Nigeria's first MakerFaire, 2012

There was the smell of burning dust and a small flash of light; I could see a scorched outlet and wallpaper. An alarm, but no fire. In fall 2012, during an early trip to Ghana, I attended what was my first hackathon on the continent—a "data bootcamp" hosted at Ghana's Council for Scientific and Industrial Research (CSIR) conference center in Accra near 37 Military Hospital. This was not an auspicious introduction. The event hosts and assembled programmers remained calm, a bit cautious, and then laughed, as they peered over the now unusable outlets that our extension chords and Internet modems were connected to. The image on my computer screen was telling: a cartoonish Robby the Robot figure was bent over and broken, and the words read, "Something's wrong. Internet connection not working."

In this chapter, I examine the discourse and practice of tactical innovation or "hacking" required by computer programmers working in Ghana during the 2010s. Like other hacking events I attended with software developers between 2011 and 2016, this workshop was typical of the visceral experience of Ghana's structural and technical woes: water, roads, electricity, and Internet were all basic challenges that coders and Internet entrepreneurs constantly had to calibrate through. But more importantly, the projects and shared visions for technology that emerged out of these events, such as the Vim! Speaker series, workshops held by DevCongress, programs developed at Soronko Solutions, and the long-running regional tech meetups called BarCamp, were all typified not only by the need to hack through disruptions but also by an overwhelming sense of patriotism and civic duty espoused by Ghana's activist-developers, endeavoring to code for social good.

Embracing a "fast, lean and agile" start-up philosophy and proposing technically relevant solutions to both software and social problems,[1] Ghana's software writers, commercial start-ups, and civil society organizers were hacking the normative conceptions of donor-aided, state-driven national development projects in this period—a practice I describe as *hacking development*. *Software development* is the term generally applied to the process of writing workable code for an application or program, iterated through a research design and prototyping process with the intention to meet some market need. The word *development* is also shared in the policies for national renewal and wealth building pursued by international financiers and NGOs such as the World Bank, which are often seen as the stewards of this work in the Global South. While the notion of development and development studies is currently undergoing succinct challenges to its racist and colonial past,[2] the term is useful here for describing the converging set of expectations and projects of Ghanaian civic activists engaged with ICT innovation for social change. I use the term *activist-developers* throughout this book to signal the kinds of community organizers, civil society advocates, and social entrepreneurs who utilize new media to disrupt status quo practices of development in Ghana and elsewhere in Africa. Within local NGOs, volunteer groups, and business enterprises, activist-developers in Ghana have pushed for civic participation and reform, primarily in the areas of open information access and public accountability. But unlike Anonymous or WikiLeaks, whose "hacktivism" in the West is often seen as anti-establishment and radical,[3] activist-developers in Ghana were not overt political crusaders. Often social elites, or "digital elites" by virtue of their entry into the tech industry, these individuals were leading and influential in Ghana's ICT community, shaping a broad notion that tech enterprise could advance social agency in the 2010s. Hacking can be defined as transgressive computer practices, including technical work-arounds, tactical engineering, iterative design, and general "making-do" in order to accomplish social and technical outcomes. In the US the term developed during the 1980s to describe computer programmers who could expertly code, solve bugs, and pull off technical security feats.[4] In Africa and elsewhere in the Global South, where digital inequality is the norm, hacking is a necessity, not simply a virtuous practice bandied about by expert code writers. It denotes agency and invention by my research participants in Ghana and can be seen as a form of innovative practice aimed at achieving technological symmetry in the world system.

In late October 2012 I attended the aforementioned "data journalism" bootcamp, hosted by Hacks/Hackers Ghana and Google-Ghana at CSIR,[5] an agricultural research center in Accra. I arrived in the morning to shadow a team of web developers and activists I had met the previous night at a GhanaDecides event in the city. The previous night's event in the Jamestown neighborhood was a raucous and communal affair, with organizers hosting a film screening and "talk back" session about Ghana's upcoming elections. In contrast, this data science–driven workshop was initially academic and staid; CSIR's scientists/managers (unaffiliated with the event) seemed more annoyed than welcoming to the young programmers and journalists at this weekend seminar, particularly given the outlet blowouts. IT developers and data-journalism advocates from South Africa and Kenya were flown in to advise the participants on Google's free data-scraping tool, Refine—an Excel-like tabulation tool used to "clean" and standardize data and datasets that would be culled from a variety of public sources. About 30 media professionals and civil society advocates had volunteered their Saturday and Sunday to learn the program, including the developers from GhanaThink, some junior developers from Busy Internet, and a team of reporters from *The Daily Graphic*, Ghana's state-owned daily newspaper.

The attendees split into teams of developers working to create apps and websites that build on publicly accessible datasets with our newly developed skills, and I joined one crew called Hack the Budget. The outcomes would be remarkable, if unlikely to benefit all attendees equally. I myself had done little web and data scraping in my technical training either as a graduate student in new media studies or in my previous life as a journalist in the early to mid-2000s; many of the Ghanaian journalists were not regular users of Microsoft Excel or any database tools at all. While the advanced programmers knowledgeably followed along as the experts explained the intricacies of database research and visual design, a solid half of the 30 participants were traditional journalists who muddled through the exercises. Despite their being adept at social media tools such as Facebook and Blogger, spreadsheets and tabulation would be a challenge. TT, a programmer and community organizer I worked with, quipped, "I wonder if people will ever use these skills. These workshops should have taught coding gradual, but now they give us all these 'refine' and fusion tables and expect magic in two days."

Most of the projects stalled when the power surge took out several of the electricity outlets we had been using to power our laptops and wireless modems, requiring most of the participants to tether Internet access to their devices through their mobile phones. Would-be hacktivists were again frustrated by the basic electrical and Internet issues of the Ghanaian sociotechnical infrastructure. On day two, Google-Ghana organizers decided to spur enthusiasm and offered an award of GHC500 (more than US$150 at the time) to the winning team, officially turning the volunteer event into a hackathon. The workshop proceeded through electrical power challenges and stalled Internet, and its participants, many of whom were organizers in Ghana's start-up community, pursued the work further. The Hack the Budget team won, and one key member, Nehemiah Attigs, would go on to transform the project into a public accountability project called Odekro. At the annual "Top Apps" awards ceremony a year later, Odekro would win an e-governance award from KACE. The app profiles members of parliament, their voting records, and their interests, and provides basic contact information for smartphone users.

In this way, among Ghana's civic hackers in the 2010s I found that the language of Silicon Valley start-up culture was used to advance an economic and political agenda for greater openness, democracy, and socioeconomic development in Africa, and many developers referred to themselves enthusiastically as hackers. *Hacking*, commonly described as exceptionally skilled software programming, remains an ambivalent word, signifying both criminality and productive civil engagement being advanced by computer programmers. *Activist-developers*, I believe, denotes a particular kind of technician in some ways specific to Ghana and its digital media producers living abroad in the diaspora. In my fieldwork and elsewhere in underdeveloped economies in the Global South,[6] activist-developers are those whose work as computer professionals, social media users, and connected civil society activists is done to advance alternative routes to political participation in what many perceived at the time as a closed political system. In a political environment in which African youth are encouraged to "sit-tight,"[7] while elders and entrenched power-brokers engage in clientelism,[8] these programmers can be characterized as social entrepreneurs—advocates using tactics from private industry and profit-minded development work in digital media to produce an alternate space of political visibility and relevance, especially for youth.

The term *hacktivist* is often used to describe tech-minded social advocates in the West who use the techniques of hacking, remix, culture

Odekro
@odekro

Check out the list of MPs who did not make a SINGLE statement in #Parliament in 4 years based on data from our #MPScorecards. #OdekroReport

FIGURE 4. A Twitter/X post from Odekro in 2016, a civil society watchdog that used publicly available "open source" data to report on politicians. Many of its members were active in #GhanaDecides and other tech-driven government reform campaigns. *Source:* Odekro (@Odekro), "Check out the list of MPs who did not make a SINGLE statement in #Parliament in 4 years," Twitter (now X), December 6, 2016, https://x.com/odekro/status/806173732511371265.

jamming, and virtual protest to advocate for more "horizontal" politics in the digital era.[9] As cultural studies theorist McKenzie Wark wrote in her manifesto on hacking, "Whatever code we hack, be it programming language, poetic language, math or music, curves or colourings, we create the possibility of new things entering the world. . . . In art, in science, in philosophy and culture, in any production of knowledge where data can be gathered, where information can be extracted from it, and where in that information new possibilities for the world are produced, there are hackers hacking the new out of the old."[10]

Hacking as such provides a new framework for who does development in Ghana. Pascal Zachary's broad and heroic notion of the "African hacker" is relevant here,[11] and that speaks to my desire to highlight the collective practices of skilled IT technicians in Ghana who utilize ICT for social good. In the 2010s well-known international hacker groups such as Anonymous and WikiLeaks shifted the discourse around hacking away from notions of profiteering and pranks that had characterized the practice in the early Internet boom toward a more radical politics. The entanglement of technology and social protest was reflected in the ways that the Occupy movement in the US and around the globe (including places like Nigeria and Ghana) articulated an anti-establishment agenda, using social media and technical communications work-arounds to protest illiberal regimes in Afro-Arab regions especially.[12] In hacktivist attacks, techniques included denial of service (DoS) campaigns, website takedowns and spoofing, doxing, malware, and other criminal, if not simply tactical, media stunts. Since the early 2000s Ghana's activist-developers, no less technically savvy or politically motivated, have mostly foregone targeting trenchant forces in the democratic republic with cyberattacks and have instead focused on the rhetoric of reform and governmental transparency that combines elements of an ICT4D agenda and "open source" information campaigns, with a push for more participation by everyday citizens in governance.

Not simply entrepreneurs or eager tech industry aspirants, many of the activist-developers I spoke with in my fieldwork were also entrepreneurs who produced apps and programming aimed at serving the public good. The most active and influential open coding communities I interacted with in this period of research, such as GhanaThink, DevCongress, and the participants at hackathons and local BarCamp events, could be characterized by both their enthusiasm for tech as a solution for economic and social inequality and, as I discuss toward the end of this chapter, an abiding Pan-Africanist sensibility in their desire to uplift the lives of Ghanaians and Africans in general. Their work continues today. In these ways, these groups can broadly be described as Pan-African futurists.

HACKATHONS: LAUNCHING START-UPS FOR SOCIAL GOOD

Many of the coders in Ghana's activist-developer community whom I interviewed and embedded myself with were active participants at hackathons and tech skills–building workshops throughout the country. A typical hackathon is a marathon computer programming session in which contestants (hackers, programmers, stakeholders) compete to develop a novel solution to an IT challenge.[13] Since the early 2000s hackathons have been used to kick-start innovation among computer programmers, corporations, and government agencies. At these daylong and all-night coding battles, convention is thrown out the window, as stubbly faced, sweaty, and mostly male programmers and engineers devise unique solutions to challenges in ways that are sometimes quick and dirty, but most of all effective. As a competition, the allure of hackathons for host companies, developers, and civic groups has been to generate a massive amount of creative thinking around a common problem or company issue. In such environments, the rhetoric of "innovation" as a business development philosophy is openly bandied about. During the early 2000s and 2010s, "innovation hubs" proliferated across the globe in places like San Francisco (I-GATE Livermore), MIT (MediaLab), and Africa—with perhaps the most well-known center being Nairobi's iHub. In Ghana, these spaces materialized at Busy Internet's cybercafe, KACE, iSpace Foundation, Impact Hub Accra, and Mobile Web Ghana, among many other offices and coworking spaces. These new sites of tech education and organizing often hosted hackathons to encourage software development for social causes.

Beyond the enthusiasm for market possibilities of new inventions and user tools, hackathons were part of the mystique of Silicon Valley and college campuses in the early 2000s—events that kept Internet start-ups grounded in the culture of computer programming grit. The contests made coding cool by immortalizing top programmers with wins for technical prowess completed in a set amount of time: 12 hours, a few days or weeks. Prizes included cash, the latest gadgets, internships, jobs, and most of all, bragging rights. Longtime computer journalist Steven Levy provided an early description of hack competitions taken to similar levels of exhaustion at engineering schools in the 1960s.[14] Open competitive hackathons, however, find their precedent in prize-winning contests held by illegal "black hat" hackers in the late 1980s and early 1990s.[15] Along with tech-oriented mini-conferences that emerged in the early 2000s, like BarCamps, THATcamps, FooCamp, and MakerFaires, hackathons have become part of the informal training for young computer programmers and web developers today. They are an opportunity to network and look for investors; they're also a chance to level up and gain fame.[16] Competitive hacking sessions received a dramatic interpretation in the 2010 film *The Social Network*, which chronicled the early career of Mark Zuckerberg. In one scene, hackers take turns completing sheets of code in-between shots of drinking alcohol at a house party. There's music and DJs and a crowd of cheering coders and lusty hangers-on. In the film, the prize for these young students is a job at pre-everything Facebook. Typically, a hacker meetup is hardly that thrilling. There are no cheerleaders. Caffeine, not alcohol, is the drug of choice for relentless code writers.

Often a sponsoring host issues an official challenge to code writers and participants prior to the start of the contest. On the day of, potential collaborators spend the early hours pitching rough ideas and mingling with other participants, who then form developer teams. Once set, the teams hole themselves up in a corner, coding, arranging, and writing through exhaustion. The competitors are prickly, arguing over fine points of programming (*HTML5 or Java? Python or Ruby?*) and are curt with outsiders. Their aim is to make a program or platform that will earn a top prize before the contest time expires.[17]

Between 2011 and 2016 I attended several ICT "capacity-building" sessions with interviewees, including hackathons, and skills-building workshops aimed at what I would term in the African context *hacking development*: participants at these sessions articulated a social agenda that used ICT for a socioeconomic development agenda, both in line with the conventional implementations of ICT and development criteria from the NGO sector such as the United Nations' millennium development goals.[18] Hackathons such as the influential "Hack Against Ebola," coordinated in Ghana, Nigeria, and Senegal in 2016, typified the ways that independent hackers, civil society activists, and government entities incorporated entrepreneurism and technology to generate social good that advances the goals of sustainability, wealth building, and political engagement in Africa and elsewhere.

Within the field of development studies, UN research associate Araba Sey has identified the two distinct ways technology has been utilized in social campaigns.[19] (1) The catchy moniker ICT4D often characterized efforts that were outcomes-focused projects in the early 2000s, advancing economic growth and integration in the global economy. These projects typically focus on getting the poorest Ghanaians and other target communities to use new technology to advance their economic self-sufficiency and improve the nation's GDP. (2) In contrast, ICT *and* development projects (sometimes demarcated as ICTD) advocate for "sustainable development" in ways that do not advance inequality or environmental degradation, in the manner that is typically associated with rapidly growing capitalist enterprises. Such projects also attempt to introduce new media appropriate to the target communities' needs and adoption strategy. The World Bank's Human Development Index is its main gauge. This field also finds the work of social scientists who explore the impact of new media on the lives of the target communities, not necessarily oriented toward an implementation goal.[20]

The academic and professional research field of information and communication technology and development (ICTD) ultimately advocates for a social transformation through technological innovation and adoption. It is a highly interdisciplinary field, whose participants range from NGOs such as the World Bank and United Nations to commercial firms looking to enter developing markets with high-impact media. Examples include the medication verification service Sproxil, serving Ghana and Nigeria, or the extension of microlending enterprises and low-cost financial services such as Flutterwave, which in 2021 became Africa's first tech "unicorn," valued at over US$1 billion. ICTD can include scholars who mix practice and research in the field as well, as evidenced in the work of Johnathan Donner in South Africa.[21] While early computer-mediated development projects focused on building media centers with desktop computers and office services, the expansion of the mobile phone industry in Africa and the Global South in general has seen more localized approaches to getting target communities to utilize ICT in their communities and work setting.[22] Sey, for instance, documents early ICTD practices in which mobile phones were introduced into communities without a stated agenda for their use.[23] Today companies such as Esoko—a spin-off of Busy Internet—and AgroCenta in Ghana provide a market service and social good by digitizing agricultural data and working with farmers in local languages in the rural regions.

The interest in ICT4D is ubiquitous in multiple sectors of the IT industry in Ghana. In 2003 the Ghanaian government established the Advanced Institute for Technology Innovation (AITI) at KACE in an effort to be seen as a leader of innovation on the continent. The center was formed in partnership with development funds from India to facilitate industry and business relationships in the ECOWAS region. International aid agencies that focus traditionally on women, education, health, rural access, and infrastructure have begun branching out, seeking IT and Internet access as a resource and having social objectives in their programs.[24] Telemedicine is seen as a panacea for rural health work, with camera phones being outfitted for diagnostic optics, and this work increasingly also utilizes drones. Strengthening access to the computer centers, access to Internet, and mobile phone networks has also been key to this agenda,[25] with the idea that target groups, especially the 50 percent of Ghanaians residing in rural areas, could reap tremendous benefits from becoming digitally connected. In all, many ICT4D projects focus on linking the poor, the marginalized, and the informal business sector to tech-enhanced infrastructures, in order to produce rapid change, with varying degrees of success.[26]

The ICT-focused social aid groups I encountered in my work, such as GhanaThink, DevCongress, and GhanaDecides, are different in that they are efforts largely initiated by local "information champions,"[27] who later seek foreign funding as they expand their projects. They also often bristled at the concept of *development* and the typical image of foreign-led development outcomes cultivated in the NGO community at least since the 1990s. Foreign aid and NGO funding is not their primary motivation for programming or social entrepreneurship, though they may seek international funding, partnership, and exposure for their work. The agendas of these civic hacker groups have emerged through their experiences as entrepreneurs and community activists in Ghana's civil society, rather than as opportunists working in the NGO sector.

In the section that follows I provide an analysis of the public discourse around tech and a description of key hacking events I attended to construct a contemporary profile of the tech entrepreneurs, online media activists, and IT-led development in Ghanaian cities such as Accra and Kumasi, which I believe are key to activist-developers' identity as civic activists.

MPOWER PAYMENTS HACKATHON: OCTOBER 2013

In October 2013 I attended the MPower Payments hackathon, named for a new e-commerce project of the successful mobile phone content firm smsGH, now called HubTel. The owners, inspired by the agility of Silicon Valley start-ups, had spent two years researching electronic payment laws in the United States and Ghana to set up an international insured and licensed online transaction firm—offering services typically provided by high-fees companies like Western Union, via novel mobile phone interface. Such "mobile money" services were only beginning to become available to the vast majority of users in Ghana at the time. Since the early 2000s and the reputation of Nigerian 419 scams and *sakawa*, much of sub-Saharan Africa lay behind a digital redline,[28] which Western financial companies and services like eBay and Amazon.com refused to cross for service delivery or commercial transactions, as discussed in the previous chapter. The platform MPower (pronounced "em-power") had overcome this barrier, with smsGH incorporating as a business in both Ghana and the US. Since 2020 the firm has signed agreements with Ghanaian banks and US financial servicers such as PayPal, which formerly lay on the other side of the digital transactional divide. Its founder, Alex Adjei Bram, started the firm as a bulk text-messaging advertiser in 2005, providing social networking and entertainment, largely enmeshed within Ghana's dominant digital practices at the time—phone-based text messaging or SMS. Bram went on to receive an executive training at Stanford University in California, and the company changed its name to HubTel in 2017. International investments helped the firm become one of Ghana's most successful software companies, with important government contracts and an annual revenue of US$64 million in 2022.[29]

But in 2013 Bram and his collaborators were still operating with the hustle energy of a start-up, indexing what communications scholar Seyram Avle states for Ghanaian tech entrepreneurs "is a way to socially and economically navigate precarious" environments.[30] For Bram and his colleagues, the hackathon was seen as a way to generate software ideas and find new coding talent. Regarding the triumph of their early transnational e-finance platform, Bram and his team joked that they drew inspiration from the film *American Gangster*, in which actor Denzel Washington cuts out the foreign suppliers in an elaborate drug trafficking scheme. "What's the line? '*I go where they go to get it*,'" they joked. Lest we conflate legitimate software development in Ghana with crime or fraud, which have come to mischaracterize computer innovation in the region, I argue that Bram and his associates' temperament reflects the tactical and iterative know-how required to develop software from the precarity of the Global South: through creative engineering and crafty business organizing within and between legal jurisdictions, smsGH at the time sought to hack out a digital solution that would be useful for the majority of users in Ghana's mobile market.

Working in conjunction with the local ICT literacy nonprofit DevCongress, the daylong hackathon was meant to attract coders who could deploy MPower in new ventures via novel smart app ideas and feature phone services. MPower itself was web based, meaning it worked primarily from an Internet site, with some functionality for mobile phones. Prize money of GHC2,000 (around USD$400 at the time) was offered for the top team, as well as an opportunity to continue developing the product with the firm. The organizers provided three catered meals, free

T-shirts and stickers, and a party at the end of the day. The contest was held in one of Accra's newest coworking spaces at the time, the iSpace—a shared office space with open and modular furniture, glass-partitioned meeting rooms, and a production kitchen. The windows and walls were painted in rainbow colors, a nod to their benefactor Google-Ghana. Google's local office, along with a UK media foundation, helped provide funds for this "social enterprise," as the company was referred to by its cofounder Josiah Eyison in 2011. On the day of the hackathon, FiiFi Baidoo, its community manager at the time and a former Google-Ghana employee, said they hoped to make the venture sustainable, with members and clients paying to rent offices in the workspace.

In 2013 iSpace sat at the top of a conspicuous six-story white building near Ako Adjei Park. (It has since moved to the Cantonments neighborhood.) A catering business was run out of the first floor, but across the street, local "chop bars" catered to commuters and men working in the area. Standing high up on the building's balcony, one could look down to see housing construction in various states of completion and disrepair and just a few shiny new businesses on Oxford Street to the north. There was an impressive landscape of Accra condos and construction cranes, but I could just as easily gaze straight into the courtyards of people who needed the most basic elements of urban development. For those living in the gutted or developing building sites, there was no running water or sewage, and garbage was piled up in alleys and forlorn open rooms. Shanties and irregular construction marked spaces between newer condominium and business developments. From another direction on the balcony, the Atlantic shore could be seen: The building sat just a few hundred feet from where nearly all of Ghana directly links to the global Internet. Five high-speed fiber-optic cables emerge from underwater, having traveled from Europe and other parts of the continent to landing stations initiating global data ports for familiar local telecoms like Vodafone, MTN, GLO, MainOne, and Expresso.

When I arrived at the contest on a Saturday morning in November, the building's Internet had been down awhile, and Wi-Fi modems were being hastily unwrapped to boost connectivity. Teams were working steadily, however. By now the mundane task of writing code and tweaking user interface was at full bore, as grim-faced young men gazed in consternation over their laptops (Dells, Acers, and more than a few MacBooks). Despite any technical glitches, the hackathon was underway. As I talked to young developers while they Photoshopped images, wrote code, and checked interfaces for quality control, it became clear that these programmers were green, still learning their craft. Most of the hackers were aged 18 to 25. Half were students or engineering majors engaged in their mandatory year of national service.

A few developers worked for the telecoms, other tech companies, and in the IT departments of banks. Some of them were first-year students at the Meltwater Entrepreneurial School of Technology (MEST). Others had been involved with DevCongress, attending free workshops on code writing and participating in mentorship sessions or open data projects. Among the 40 contestants, only 2 were young women. Such youth was the hallmark of the developer scene in Accra at the time. In my fieldwork with IT developers in Ghana since 2011, and in networking at trade shows, policy conferences, app awards ceremonies, hackathons, cybercafes, and other IT centers, I rarely met a code writer or software engineer over the age of 30. When I did, they were typically senior developers and managers with established firms.

Later during that visit to Accra in 2013, I also visited Herman Chinery-Hesse's firm theSoftTribe, and the age gap there seemed inescapable. Chinery-Hesse, as discussed in chapter 1, is a charismatic organizer, programmer, and businessman now in his sixties. At the spartan mansion that serves as the developer home of his software company, most staffers are typically a good 20 years younger than he is, from the receptionists straight through to the project managers, while the firm's principals are a bit older, in their thirties and forties, serious and suited. At the time I met almost no programmers with more than 10 years of experience. Among the older software developers there were a few who went to engineering school; the majority had migrated to this field, learning code languages and networking protocols along the way.

HubTel's Alex Bram initially studied chemistry in college. In his thirties then, he said his age was always an issue in business meetings in the 2010s with venture capitalists, banks, and US and Ghanaian government officials. "We have no old guys on our team," he said to me at the time. He describes a dispute in front of a city judge dealing with a zoning complaint for his business. The first thing the judge talked about was his age. The magistrate was shocked that "the youth" could be responsible for this two-acre block of construction, with a pair of five-story office buildings. Having graduated from Kwame Nkrumah University for Science and Technology (KNUST) in the early 2000s, Bram and a fellow science major decided to become web entrepreneurs, teaching themselves code and going back to school for business degrees. His partner, Kwadwo Seinti, the company's chief technologist, studied electrical engineering in college. His education as a computer scientist was largely informed by free classes and lectures offered on the Internet. During a lull in the competition, he showed me the syllabus and videos of an MIT professor he was particularly fond of. He and another programmer argued over who held the best online machine learning courses, Stanford or Harvard.

Despite several mentions of Microsoft mogul Paul Allen and Apple cofounder Steve Wozniak, this hackathon would not likely qualify as a stereotypical corporate or civic tech fare. Instead of Danish pastry and a boxed lunch, caterers set up an impressive array of local foods served by caterers: *banku* and tilapia, jollof rice, fried plantains, crab stew and fufu. In the spirit of the charismatic self-development exuded everywhere in Accra, an earnest motivational speaker came to advise the young programmers how to effectively conduct themselves as an entrepreneur (with "passion and faith"), while the judges deliberated the winners. The intermission then included cocktails and palm wine, served out of coconut shells by a drink mixologist. A DJ played a 45-minute set of dance music, while the chief technology officer and a few others hopped around to hiplife and azonto music. (Maybe this was *The Social Network* after all, I mused to myself.)

Still, what set this app contest apart from the hackathons one might find at a developer event in Palo Alto before the advent of "tech bro" culture in the late 2010s was the uniformly social impetus for most of the teams' final applications, significant for an event squarely aimed at building commercial payments tools. Of the seven projects, the majority were social welfare–oriented apps, attempting to bridge multiple technical and social divides. They included 'Frika, "an African crowd-funder" for large social projects like sanitation in rural villages or seeding funds for small firms. A separate donations app was produced by one team, intended to support key "life events" in Ghana—gifts and money for naming ceremonies, weddings, and funerals, important activities that often leave out Ghanaians unable to travel within country or useful for those living in diaspora. Nearly all the projects incorporate SMS-based payment and text interactions, attempting to reach the many users in the Ghanaian market who connect via feature phones—the slim number-pad-driven cellular devices sometimes referred to as "candy bars" or "yams." For users operating in these zones of a digital divide, this is a common trajectory of tech use—reliance on SMS mobile communications, low-bandwidth Internet usage, Windows rather than Apple hardware-oriented computing, Android mobile apps, open source tools, and free (at times illegal) software exchange. At the time of the MPower hackathon, as

FIGURE 5. A developer explains his code during a pitch competition at an MPower Payments Hackathon, Osu neighborhood, Accra, in October 2013. Photo by author.

smartphones were only beginning to penetrate Ghana deeply, the majority of high-tech phones worked off of the Android mobile platform, owing to the program's openness and flexibility in its product development and pricing.

There were only a few commercial projects: an aggregator portal for consumers seeking online deals from local businesses and an Expedia-like local fare hub featuring Nigerian, Kenyan, and Chinese airlines. One app fashioned a Bitcoin payment system to MPower along with a Bitcoin generator. This was novel, as cryptocurrency was hardly as well known in 2013. Caridad and two other donation apps were designated the winners in the end. But in another novel experience for Ghanaian hackathons, HubTel agreed to give the other contestants GHC500 toward their projects if they finished by end of the week, and it extended to the losing competitors opportunities to continue developing software with the tech firm in other ways.

Considered in one way, hackathons in Ghana represent a kind of high-tech localization of the tech subculture of Silicon Valley. The innovation and practices here also demonstrate a particular valence of tech practices that link Ghana to a kind of global user, one whose practice is particularly dominant in the underdeveloped economic markets, and for peripheral netizens in the Global North, in low-income urban neighborhoods, and some rural communities, where limited or metered access to Internet is the norm. The website for OasisWebsoft, whose CEO Raindolf Owusu was an active participant at these meetups at the time in Ghana, states, "We are committed to building infrastructure that will ensure that the West African sub-region is not left behind in the continuous evolution of information technology."[31] Hackathons and other tech events reveal a network of tech innovators and a geography of tech development in the capital Accra, making physical sites and real-world events such as these important, despite the immateriality of tools such as software and online content.

Many who look for technology in Africa present the continent as a sort of digital safari. Interlocutors routinely search for "authentic" tech innovation on the continent, as if tinkerism and improvisational bricolage were the only sciences available, a narrative one might read in blogs like Afri-Gadget.[32] Started by iHub's founder Erik Hersman—a White American active in ICT development in Kenya since the early 2000s—the website profiles African tinkerers, local inventions, and mechanical improvisation. The blog attempted to highlight for the world the skills and, more importantly, desire of Africans to achieve some sort of technological symmetry with Western materialism. AfriGadget, now no longer active and mostly archived online, purported to demonstrate how Africans are "solving everyday problems with African ingenuity." The site featured stories on homemade helicopters, cobbled together with scrap metal from roofing, as well as articles on biodiesel engines using manure to power villages. It is a project that attempts to do a service but that in many ways reinforces the foreign gaze for an audience of geeks and technocrats. With the Other firmly established and reinforced, these outlier narratives attest to singular cases of African invention, but one interpretation might be that despite these creative ventures, the persistence of poverty on the continent seems natural and its problems incapable of being remediated. At the same time, rather than acknowledging ongoing asymmetry between the West and the Global South, these peculiarities distract from the broad needs of development, infrastructural and economic reforms that could pave the way for more innovation and openness in places like Ghana.

With AfriGadget as its filter, the African techno-voyeur renders even the efforts of developers at hackathons and networking events exotic, or as Jenna Burrell describes, "invisible." Thus there is an impulse echoed in the anthropology of innovation to look for technology in Africa that speaks to both the "primitive" and the highly technical, while ignoring the making of modernity inherent in these projects.[33] This impatience with the foreign gaze was evident during an interview I had with Alex Bram, when my line of questioning around tech innovation pushed for something more than SMS and Android applications. "All you guys from the West want to come here looking for the 'African technology.' . . . It's not [in] some device made from rusted metal and springs," joked Bram that day.

HACKING AS PAN-AFRICAN

A key figure among activist-developers I encountered over the years in Ghana is Jemila Abdulai, who now works as a development consultant and digital strategist in Accra. Abdulai had experienced Ghana unevenly in her youth. The daughter of a computer programmer, she was born in Nigeria; Her family was part of the brain drain that left Ghana during the Rawlings military regime. She attended school in Norway, then returned for secondary school in Accra. In diaspora in each of these places, she felt exclusion for being Ghanaian, Black, a northerner, and Muslim, respectively. "It makes you very cognizant of the minority story, the story that's not mainstream or not the status quo. And so I am very aware of how easy it is for those stories or voices to be sidelined," she told me in an interview in 2014.[34]

I first encountered Abdulai as one of the key faces of GhanaDecides, a civic engagement campaign started during the 2012 election, in partnership with the online writers' group BloggingGhana and the STAR-Ghana development project.[35] Abdulai was studying international economics at Johns Hopkins University-SAIS at the time and preparing a master's thesis on the impact of the Internet on politics when momentum started building with the youth-focused campaign. She pivoted from her studies and became a volunteer with GhanaDecides, traveling back home to participate on the ground and offering her skills as a social media influencer while living abroad. During the 2012 campaign she remotely managed GhanaDecides social media accounts while at school, working night shifts when workers on the ground were asleep.

A few years after that election, Jemila participated in a series of social media influencer interviews, asking Ghanaian content producers how they got their start. When asked to complete the phrase, *Being Ghanaian means to me* . . . , Abdulai stated:

I believe in my country. I cherish our history but realize that in order for us to create a new future, we can't just rely on what has always been. We should be open, but at the same time we should also find the value in our past. Sankofa—basically go in to the past and take what's relevant. I feel a certain pride in knowing that Ghana is one of the first countries to gain independence in sub-Saharan Africa, and was a pioneer in many areas. But I also feel that we have a certain sense of responsibility to uphold that.[36]

As an activist-developer, Abdulai's virtuosity in using digital media for civic engagement marked her among the lead users of content production via her blog Circumspecte.com in Ghana during the early and mid-2010s. With a focus on women entrepreneurs, and providing advice for start-ups and freelancers, she didn't limit her focus to the subject of traditional development studies concerns, such as health or basic literacy. Instead, Abdulai became what Wilson and Wong described in an early account of ICTD research as an "information champion," training nonprofits in digital content production, business development, and other forms of professional communications.[37] She told me in our 2014 interview, "I am just a creative, and I use different mediums. . . . I am someone who self-educates when it comes to tech. . . . I believe once you learn something, I don't think another person should have to suffer to learn it. You have the information, I believe information sharing is important. One piece of information can change someone's life."[38]

Thinking about Jemila and other Ghanaian digital media activists as hackers rather than as simply community organizers or tech entrepreneurs helps us identify the disruptive ways that digital media are being deployed for development in the Global South. Abdulai graduated from SAIS and began to work for the African Development Bank, traveling throughout, to places like Côte d'Ivoire and Tunisia, but she made her home in Ghana once more. As a development professional and blogger, she had been posting online stories and videos since 2007, featuring stories on emerging women-owned businesses, tips on freelancing and working as a social influencer, and posting images from the many African cities she ventured to professionally. Reflecting on her own experiences as an entrepreneur and creative, Abdulai stated she wasn't satisfied with the image of Africa being projected in the development field, and even less with the stakeholders in the room. In her free time, she began to use Google+Hangouts to host roundtables with African and Black diasporan development professionals, both to build camaraderie and to discuss policy on a range of issues from social media and privacy to elections and World Bank initiatives.

In Google's first iteration of Hangouts, the Internet application was intended as a competitor to Skype, allowing for easy VoIP capabilities, and in the African context, allowing for the kinds of transnational social interactions that Mantse described in chapter 1. Redeployed as an organizing tool by Abdulai for GhanaDecides and with other development-oriented activists, the application enabled connections that may have otherwise been stymied by in-person social networking decades earlier. She said in 2014, "In my experience in Washington, DC. You have all these development experts talking about African issues, but most of the time you don't actually have Africans talking about these issues themselves. The panels always tend to be largely non-African. I always found that interesting and somewhat unsettling."[39] Abdulai's sentiments, along with other statements by activist-developers throughout my fieldwork, indicate that such modes of hacktivism serve not only as a means of overcoming information gaps and other digital divisions, but also as a means of empowering African sociopolitical consciousness and ultimately a sense of collaboration and unity inherent in a Pan-Africanist sensibility that Abdulai often invokes.

In Gabriella Coleman's pioneering ethnography on the hackers involved in the creation of the Open Source movement in California, the media anthropologist documented the ways that hackers celebrate freedom of speech, irreverence for authority, and a libertarian ethos of free enterprise, which she frames in the Western context as "possessive individualism." Coleman states, "Their passionate commitment to hacking and especially the ethics of [open] access, enshrined in free software licensing—expresses as well as celebrate un-alienated, autonomous labor—also broadcasts a powerful political message."[40] Coleman also identifies a mindset that hackers embrace, that is in many ways directly related to the sociotechnical systems of life on the continent. She states that "hacking, whether in the form of programming, debugging, running or maintaining systems, is constantly frustrating. . . . In encountering obstacles, adept craftspeople, such as hackers, must also build an abundant 'tolerance for frustration.'"

In the epigraph of this chapter I quote the words of tech developer Omoju Miller, a colleague of mine who in graduate school at UC Berkeley worked to bridge similar tech enterprise in her home of Nigeria with software firms in the US. Now an industry executive, Miller described the iterative craftsmanship required to work through infrastructural breakdowns, lack of materials, and general uncertainty as being the hallmark of any enterprise on the African continent: "We know how to hack things," she said to a developer audience back then. This iterative or "make-do" attitude of any developer, especially those working in the Global South and periphery, names a practice of hacking. However, the ethos of hacking as a creative or generative practice meant to overcome system impediments is not confined to the technical strategies of computer use. Put in the service of socioeconomic development and society, such tech-focused social entrepreneurs in Ghana and Africa generally could be said to be engage in hacking development, especially since the mission of such groups is in accordance with the ideals undergirding the UN's agendas, such as the sustainable development goals (SDGs).

The notion of hacking as a social strategy and political ethos has taken root in Ghana as it has among other programmers in the West—although most activist-developers in Ghana were interested more in reform than in insurgency. Ghana's digital elites are among the biggest proponents of social action, using IT as media of change. Once more, these ventures embrace a nonhierarchical and tactical approach to participation and strategy, in contrast to the top-down modernization policies of the postindependence era, or contemporary ICT4D-focused projects from the government and donor agencies such as the US Agency for International Development or UK Aid Direct. But these efforts by activist-developers are not strictly capitalist, market-driven ventures. Many of these Ghanaian entrepreneurs believe that their businesses are doing the work of traditional economic development. Frustrated with the global image of Africa's abjection, the rhetoric of the anti-establishment, fail-fast, "light and swift style," and DIY pose of Internet entrepreneurship appeals to many of these reformers.[41]

In my fieldwork, this was most evident at the phenomena of organic tech meetups called BarCamps, run by the nonprofit group GhanaThink and headed by Ato Ulzen-Appiah. The original BarCamps started as a loosely organized conference of web developers in Palo Alto, California, in 2005, in the heart of Silicon Valley. In opposition to rigid and professionally produced industry conferences and exclusive tech meetups such as Tim O'Reilly's FooCamp, organizers at BarCamps opted for an organic itinerary, democratically deciding upon panels and presenters in the days leading up to the event or even sometimes on the same day. It was generally expected that participants at BarCamps must either present or assist with an presentation; no passive observers were allowed. This gave the meetings of network engineers, computer programmers, salespersons, and other tech enthusiasts an of-the-moment, emergent energy, which supported generally cutting-edge topics and themes.[42] In

Ghana, GhanaThink has run BarCamps throughout the country since 2009, keeping the events open access, no cost, and free form as part of their overall mission to engage youth and tech developers in social entrepreneurship. The first event was held in Accra and the second in Palo Alto on Stanford's campus, as DiasporaCamp. Every year since, GhanaThink has sponsored 8 to 12 meetups held throughout all of Ghana's major municipal regions at colleges and community centers. As mostly mentorship sessions, the events feature social influencers and start-ups who share their personal journeys as professionals and solicit collaborative projects.

In 2016 organizers from BarCamp created the initiative National Volunteer Day (#NVD), revitalizing Kwame Nkrumah's birthday, September 21, as an annual service holiday. At an event I attended in 2016 in Sunyani, Ghanaian developers led senior high school students in Wikipedia editing sessions, promoting the idea that Africans needed to self-publish more media content about themselves online. Tigo, one of the five major telecom operators at the time, announced the rollout of its live-birth registry tool at the event, describing a partnership with the National Health Service to increase the number of birth certificates via mobile devices. The day began with organizers leading participants in the Ghanaian national anthem, "God Bless Our Homeland Ghana," with its stirring independence era lyrics: "Raise high the flag of Ghana / and one with Africa advance; / Black star of hope and honour / To all who thirst for liberty."

According to Ulzen-Appiah, the tech meetups are meant to instill a sense of resilience in young developers working through persistent infrastructure challenges such as #dumsor and to provide for models of "excellence" and achievement coming from Africa. In an interview in 2016 he gave me the example of a BarCamp app maker who transformed the traditional marble-and-board game *oware* or mancala into a mobile game, playable via smartphone. Mancala is one of the oldest games in the world. Logical and mathematical, it requires players to collect pieces (shells, stones, marbles, etc.) in small bowls indented on a board made of wood and jump your opponent in a turn-based race to the finish. A virtual version of the game was novel at the time. Such gamified African cultural expressions have been advanced by other BarCamp mentors, individuals such as Eyram Tekyiwa, cofounder of Leti Games, who also has presented at the conferences for years. His software titles include an open source DJ interface, a street football game for Facebook; a KandyKrush clone using African superheroes; and iWarrior, a vertically scrolling shooter game.

iWarrior is significant not for the game play but for the theme and aesthetics it incorporates. A Masai warrior "bushman" in full battle regalia (skirted, mudded hair, etc.) fires arrows and spears at animals stampeding through his village. The environment and ecological dangers of life in the semiarid desert are immediately brought into play. African cosmology is present as well, with spirit forces invoked through thunder and ominous voices, affective media in the game's interactive design. While a bit essentializing, the game exposes players to a degree of familiarity with African rural life not typically rendered in gaming at the time, save for negative depictions in titles like *Resident Evil.* "We have a lot of the conversations like, 'You can do this yourself. You can learn to build games, here are the tools.' Look at this cool thing that this other guy has done with technology. He has taken [mancala] that we normally, we play at home and he's put it on Android, he's done it. He made a 3D version of it. Their eyes just go wide with excitement." The Pan-Africanist ideals espoused by Abdulai and at BarCamps are captured succinctly in GhanaThink's mission statement from its website in 2013: "We are an Africa-focused think-tank based in Ghana. Our mission is to mobilize and organize talent for the benefit of Ghana and the world. . . . GhanaThink seeks to mobilize and deploy human capital for the primary benefit of Ghana, and consequently of Africa and the whole world, for we are each our brother's keeper. . . . GhanaThink thrives on the free flow of information. . . . Our principles are motivated by this higher purpose: To restore the confidence of our peoples, especially the most disadvantaged. We believe in homegrown solutions to problems, because these are inherently more robust."[43]

SWAGGER NEEDED

In their most productive capacity, these developer-driven tech meetups are demonstrative of the sophistication, ingenuity, and business savvy of young Ghanaian computer programmers who position their professional trajectory alongside the start-up culture of Silicon Valley. For my purposes, the hack events I attended also represent a ritual of tech sophistication and, at times, a form of acculturation, as if to signal that appropriation of certain IT practices and tools could also be portrayed as achieving a kind of modernity imagined as "Western" and "developed."[44] The actions and rhetoric of these tech practitioners advanced both sensibilities of a local, "native" technoculture as well as a desire for modernity, given the material conditions of exclusion and breakdown in the postcolony. At the MPower hackathon, despite the localization of designs for geographic and technical specificity, what was striking and disappointing about the icons of the winning programs was their decided lack of Ghanaian or African presentation. Developers mostly cribbed images of White Westerners from the Internet for their marketing. All the interfaces were written in English, rather than in local languages. (One app was curiously labeled Caridad, Spanish or Portuguese for "charity"; while the developer explained the origin of the name during his pitch, he didn't give a reason why he chose that name.) Perhaps these aesthetics reveal an understanding about the developer's intended user or further reflect a self-deprecating understanding of Ghana's position in the global tech market and the perceived benefit of software coming from the West. Many of the programmers I talked to that day, echoing sentiments I heard elsewhere in Ghana, stated that the American or European tech services were inherently better in quality, more profitable, and more secure. Even for some patriotic programmers it seemed, the West itself was synonymous with technology. One developer at a hackathon commented to me, "We Africans hate ourselves. It's not just that we don't like Nigeria or Ghana, Ga or Ewe. It's because we've been conquered. Some of it's just shame."

Some companies I encountered at the time boasted brand names that seemed African but were not ethnically or linguistically specific. For some, as was explained to me, it was important that their products *not* be seen as Ghanaian, carefully burying information about the developers on products' websites so as to seem American or European, or at least South African or Kenyan, whose digital products were presumably more respected.[45] For me, this furthered the ways that in the West the tech industry configures its users as both White and affluent. As a seasoned project manager remarked during an interview, "We have lots of software developers [in Ghana], but customers still demand software applications from outside. . . . They always want to know if it's Microsoft or Cisco. They don't trust their own."[46] The effect was that Ghanaian code writers would not typically market software platforms that they had devised themselves, and that local innovation was spurned in favor of notions of an "industry standard," the hallmarks of which are established by Western product design and branding.

This led to an outwardly focused approach by some start-ups at the time. At the MEST, I interviewed developers at RetailTower, who told me that their product was not primarily for Ghanaian consumers. They were trying to build an e-commerce platform for users in the West, an alternative Amazon, that would also be useful on the continent, where the e-commerce giant was only doing limited business with users who held American bank accounts. The tactic led to some success: DreamOval, RetailTower, and Nandimobile (see chapter 1) had won multiple awards at events such as LAUNCH, an international start-up competition based in San Francisco, and at the World Summit App Awards in 2013, alongside mobile payments app InCharge Global. DropiFi, a business started by Meltwater students, was picked up by 500 Startups, a Mountain View, California–based venture capital firm. In an interview with CNN that year, the founders acknowledged that though their customer relations software had been developed for Ghanaian businesses, their clients would most likely be found abroad. "Currently we are focused on the US and international market—the US, UK, Canada—but in a couple of years we want to become SaaS leaders in Africa."[47]

While the Kenyan mobile payment system M-PESA has often been touted as a vernacular and democratizing technology for Africa's infrastructure and economy, the success of RetailTower and some other Ghanaian software applications also marked the differing ways Ghanaian digital elites and the networked majority interacted with digital media. Smartphone usage was in the minority until the 2020s, and while the lack of landlines and in-home Internet continues to be a problem for the majority of Ghanaians, digital and social elites enjoy more consistent access to high-speed broadband services. Even in the space of engaged civic activism, such as GhanaDecides' broadly aimed voter participation campaigns, there was a clear consumption divide, and differing class modes of tech habitus became evident. The group's videos on YouTube at the time had a few hundred views, in contrast to the thousands of clicks on Ghanaian music videos, Ghallywood/Nollywood films, or evangelical Christian media content on the same platform. Such divisions illustrate the heterogeneity of global technoscapes and, at times, the limits of elite power. "In terms of Internet users, we [developers] are hardly the majority" said Abdulai in one of her Hangout sessions with other African developers.[48]

At times the social divisions between developers who are locally focused and others who design apps and programs based on usage patterns in the West seemed especially prominent. In our conversations, Herman Chinery-Hesse, the veteran developer, bristled at the model of foreign-trained or incubated programmer talent at nonprofits and at NGO-led endeavors such as the British Council's annual "Duapa," a social good–focused app competition.[49] "Innovation will come from Africa, not the other way around. Other guys have a school, I have a development firm with commercial platforms. For a first-time developer here in Ghana, I wouldn't let them touch code. I'd sit them down and have them listen to a bunch of Jimi Hendrix and watch crazy movies to think creatively. For me learning how to think as innovator can be likened to be breaking the mental chains of slavery. One needs swagger to be a coder here."

INNOVATION BEYOND TOOLS

The use of ICT for development is an imperfect practice, and for many, ICTD remains ambivalent, as it is sometimes posed as problematic within the research and development studies space.[50] Ghana's activist-developers approach ICT and development in myriad ways. While critiques of "governance," corruption, and nepotism inform a familiar notion of dysfunction in African politics, what the critical anthropologist Jemima Pierre refers to as the "racial vernaculars" of development discourse,[51] the narratives of activist-developers are typically "apolitical" (their terminology), aimed at energizing participation in elections and accountability in civil affairs. The rhetoric of the "trade not aid" movement and a desire for self-reliance are reflected in the DIY ethos of programmers and developers, as evident in the tech industry throughout Ghana.[52]

During interviews with contacts, in blog posts, in Twitter/X messages, and in other discourse online, Ghanaians routinely expressed dissatisfaction with the government's efforts to build job opportunities and maintain infrastructure like roads and electricity. Tech-minded advocates were also dismissive of traditional aid NGOs as being politicized and corrupted or continuing African dependence on the West. One social enterprise manager working at Impact Hub Accra said to me:

> I think there's certainly situations where aid is needed, in a humanitarian sense, so like the Syrian refugee crisis needs aid. It looks bad if NGO project targets are not met, therefore, more money gets allocated, but problems are not solved. At all different levels, from the village, to sitting at the table with officials, the whole sector is skewed in various ways. The incentives are all messed up. It's much easier for the government and the bilateral agencies (e.g., USAID) to sort of fudge the report to make it seem as if the government meets its targets even if they haven't by some measures. It's politically more difficult to go back to Canada or the United States, or Netherlands or Germany and actually tell those governments, Ghana did not perform and we should not release funding. In that way there's no real incentive to perform in this sector.

Akua Akyaa Nkrumah, a Ghana-born, US-educated environmental engineer I met at BarCamps in 2012 and 2016, spoke to me proudly about her efforts to produce a self-sufficient social enterprise in Accra. "I love trash. I call myself an eco-entrepreneur these days," she said. We discussed the emergence of the "upcycle" movement in the US—apps and retailers that purchase used clothes and furniture to repackage them for the thrift market. "That's cute," she said to me then. "How many bow-ties can you make from curtains? I am thinking in tons. I can walk into a room and tell you how many tons of trash [are] being generated."[53]

Akua Nkrumah started *Green Ghanaian*, a blog and consulting business, after hanging out at BarCamps for many years. Her full-time job was working for a local refuse company. Under her watch, Green Ghanaian led volunteer environmental cleanup events throughout the Greater Accra region, and she worked with communities to do "citizen science," including testing of water ways and urban spaces as part of their public advocacy. The synergy between her consultancy and her employer Jekora Ventures, a trash removal service, helped them both become signature presences in Accra during her tenure. Active on social media, she promoted clean tech and tech-led government accountability in the waste industry. She posted on Facebook in 2014:

> We have 2 customer service lines where clients constantly call in to complain when there's a break in service, when pick up is late, when prices increase. . . . We accept all calls, listen with deep concern and address complaints as soon as we possibly can. All I wish is that such call lines were available for our government. . . . Real places where people could call and voice their dissatisfactions and be spoken to by real people. But that would imply that our leaders actually want to solve our problems, which I doubt they've ever even considered.[54]

Speaking to a group of youth in 2016 at Ashesi University, she positioned herself as a capitalist, stating that the market could and should provide solutions to Ghana's social problems: "Let us compete amongst each other for doing good. I'm trying to do a good thing. I'm trying to beat you at doing that good thing. It makes the whole society better."[55]

Other activist-developers I met, however, balance market-driven enterprise with NGO-driven support for development initiatives, utilizing foreign funding and interests to help propel their local projects and self-articulated efforts at economic and social growth. In 2012 I visited Hub Accra, a small coworking-space for programmers and start-ups located in a sweaty, single-floor office building across from the dusty football court at Ako Adjei park in Osu. An upper level of the building, still under construction, sat with open steel girders and unfinished walls. Inside, I met aspiring tech entrepreneurs selling their services in web design and

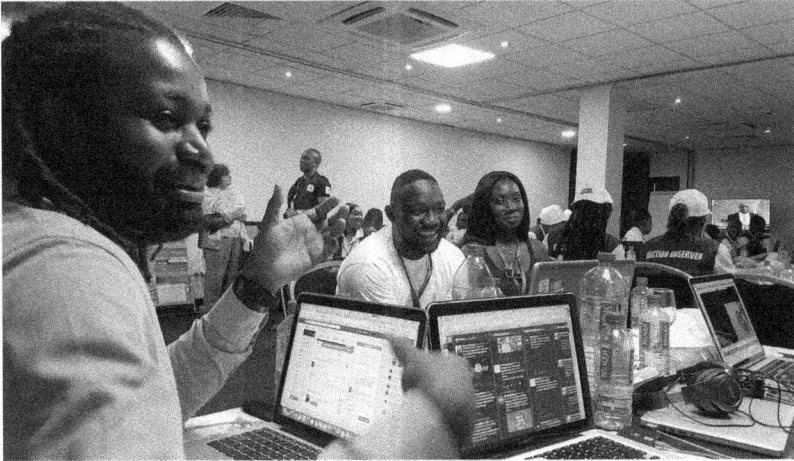

FIGURE 6. Fred Frimpong volunteers with members of the civic group GhanaDecides at an election observation event as they track news developments and encourage voter participation via social media in December 2016. At the time, Frimpong operated a financial tech start-up called Fido. Photo by author.

network engineering to local ad agencies and law firms. The scrappy organization was also the local headquarters for the Open University of West Africa and a new business management firm called Growth Mosaic. Visiting a few years later in 2016, the coworking space was now called Impact Hub Accra, having joined a worldwide network of tech incubators who in their words serve as a "catalyst for entrepreneurial action," and was receiving development funds from the US Broadcast Board of Governors, the British Council, Facebook, Siemens, and others. The organization moved into a sleek 1,700-square-foot facility in the same neighborhood, a building filled with glass windows, a makerspace with heavy machinery, film production studios, and private offices. Caterers served tilapia and espresso out of kiosks in the courtyard.

Google-Ghana has long been a partner with the new Impact Hub, and the site became one of the earliest locations to receive the company's high-speed fiber-optic network in the city, partnering with its Internet backbone spinoff CSquared. By 2016 Growth Mosaic, Impact Hub's anchor client and a partner in its business development, had become a "fully mature" B-corporation with several employees and more than 20 clients for its social enterprise mission—investing in small firms with "slow growth" investors and donor aid grants. These firms or social projects included some ag-tech start-ups, a rural waste disposal company, a midsize shea butter distributer, and a tech-skills training firm. This model of development, its managers argued then, could perpetuate economic and social growth in traditional areas served by NGOs, but also could spur profit by using business models, a practice referred to as *social entrepreneurship* in the development field.[56]

Still emerging from these new investments in 2016 when I spoke with him, William Senyo, one of three cofounders of Impact Hub Accra at the time, said the center's operating costs were 60 percent covered by external grants and 40 percent from membership rents and ICT services rendered to clients. On the organization's website, the colorful iconography of the UN's SDGs is prominently featured, with its emphasis on 19 key areas such as poverty, hunger, and "innovation and infrastructure."[57] "This is a social enterprise," Senyo told me. "We care more about impact than we do the bottom line. We are just getting to [be] sustainable, but we need grants to survive."[58]

Reiterating a comment I heard from other activist-developers who ran their own businesses in Accra, Kumasi, and elsewhere, Impact Hub's operations manager, Victor Kelechi Ofoegbu, joined our discussion. "Almost anything you do in this ecosystem will have deep social and economic impact. We're in postcolonial Africa. GDP is stagnant. Daily wages rarely peak above $2," said Ofoegbu, a longtime resident of Ghana, whose family is originally from Nigeria. "Nothing works the way it should. Schools don't serve us. Governments are absent, and the infrastructure is failing. Startups are stepping into the void to solve these problems." Senyo later echoed the sentiment: "Nothing the [government] does can contribute to innovation: When they tried with education, by laying universal (rather than local) standards . . . [squashes his hand in a flattening gesture]. Negativity! We are terrible in science, technology and mathematics. The government wants to create data centers. Amazon web services is more than enough for me. How is that a strategy for national development?"

Moderating a bit, Ofoegbu chimed in, "But we can't just innovate our way out of this." The pair talked about their growth as activist-developers over the years, realizing the necessity to work with government officials both politically and tactically. They describe efforts to help the local mayor's office in Accra hire a CIO (chief information officer)— an administrator whose office would provide access to municipal data. Such "open data," Senyo argued, could be used to solve problems associated with infrastructure, potentially to start new businesses, like Akua's Green Ghanaian. Senyo said, "I've found, hacking things for a long time, that change is incremental. I've always thought local and small. . . . [But we] think national all the time because we were focused on national liberation. Now after all these years . . . we see that not much has changed. We need to think about economic emancipation."

The comments of my interlocutors and interviewees mirror in many ways an oft-referenced 2007 TED talk by Ashesi University's founder Patrick Awuah. Arguing for the need for liberal arts education in Africa, Awuah had a career as a senior program manager at Microsoft before starting the college that year. In this talk, Awuah states that the quandary of Africa's development woes falls into three key categories: "Free markets, rule of law, and infrastructure." He argues that solving these issues is the central mission of his liberal arts and innovation-focused college in Ghana: "What Ashesi University is trying to do is to train a new generation of ethical, entrepreneurial leaders. We're trying to train leaders of exceptional integrity, who have the ability to confront the complex problems, ask the right questions, and come up with workable solutions."[59]

On the ground in Ghana, configuration of the network society seems at times hardly as egalitarian and democratic as the grid imagined by Paul Baran and others. In the period since 2018, when the majority of this fieldwork was conducted, with the pandemic and financial slowdowns becoming more pronounced, the promise of Africa Rising began to fade following inflation, rising fuel costs in Ghana, debt downgrades, and International Monetary Fund (IMF) interventions. Currency devaluations have made Ghanaians less enthusiastic about the role of entrepreneur-led social change. Activist-developers paved the way for civic innovation, but without robust development of pro-Pan-African enterprises rooted in social and democratic outcomes, ICT4D has become a less robust project. As I explore the specific sociotechnical work-arounds in diaspora in the next chapter, disjuncture in Ghana's digital mediascape becomes more clear. But the generative work is this: with an uncertain market future, Ghana's activist-developers are proving to be immensely creative and resourceful given the continuing woes of a developing economy with a tenuous infrastructure and enduring narratives of Africa's technological anti-modernity. The notions of DIY and self-reliance are not only meant to stunt the continued system of dependence on institutions such as the World Bank and what activist-developers often referred to as "dead aid."[60] Technology and social entrepreneurship for social actors seemed rather to be a *systems exploit*, to use the language of hacking: an iterative and creative work-around utilizing a system's own resources, in this case neoliberal models of capitalism. These tactics reveal themselves as

FIGURE 7. Impact Hub Accra facility in central Osu in 2016, offering coworking, film production studios, a heavy-machinery maker space, and a direct access point to high-speed urban Internet backbone. Photo by author.

key resources for Ghanaians who would meet the very real need for economic growth, political transparency, and civic participation in an increasingly connected yet frustrated democracy. Taken together, the efforts of Ghana's activist-developers help to typify a broader liberatory ethos of African digital innovation and a Pan-African zeal. If the structures of the Internet and global ICT can be shown to be as limiting or controlling as they are enabling, a decolonial Pan-Africanist futurist disposition with regard to techno-modernity reflects the high ideals of African liberation. In the 1950s and 1960s, nation-building efforts after political independence were also intimately focused on material interventions, especially in Ghana, where industrial and state projects were focused on energy generation and the development of state-driven start-up industries.[61] In these ways, the Pan-African futurists documented here similarly invoke a materialist disposition, focused on tangible connections, reclamations of culture, control of commerce, and redevelopment of natural/indigenous resources and wealth for the benefit of Africans.

CHAPTER 3

Digital Diaspora

CIRCUITS AND PRAXIS

Mantse reaches into an impossibly narrow closet in his bedroom apartment in the Lake Merritt neighborhood of Oakland, California, to retrieve a bulging, black trash bag.[1] He smiles, showing me what's inside: over two dozen mobile devices in various states of disrepair and abuse, including flip phones, phones with small plastic antennas, scratched-up Razrs, and small Nokia and Ericsson "yams." Mantse sends these and laptops back home and sells them on the secondhand market for cash on hand in Ghana or to give to relatives. "All my used tech stuff, I normally ship it back home. Printers. Every time my printer junk, I just pack it up, and when I'm doing shipping, I send it back home. I buy a printer for $150. It costs me $100 bucks to fix, so I buy a new one, and mostly the newer ones go faster and it's cheaper. So, I buy a new one and ship the old one back home."[2]

In the early 2000s Mantse was a well-known presence in the Bay Area's Ghanaian community, especially in "youth" circles (anyone under 40), as he would regularly blog and send updates via email to others in the African migrant communities. He was also a DJ and promoter, which gave him cachet with the young adult crowd. In 2008 he received a "Youth Leadership" award from Ebusua, one of a handful of independent benevolent groups in the Bay Area associated with Ghanaians in Northern California. Mantse was one of the first people I interviewed on the topic of digital diaspora in 2009 in California. At the time, Facebook was still emerging, yet to become the world's most dominant social networking service. Smartphones and high-speed broadband were just becoming widespread in the US. In the early 2000s smartphone penetration for even emerging "lower middle-income" economies in Africa numbered at under 20 percent of all mobile phone users.[3] Connecting back to Ghana digitally took work. For diasporans like Mantse, tools such as blogger networks ("webrings"), listservs, and alumni message boards served an important function in finding news and debating current events, as well as negotiating shipping items back home. Mantse first started using the Internet in the early 2000s; as part of assignments for college work, he was required to go online to do research and submit class papers. He hadn't been exposed to the Internet much as a teen in Accra during the 1990s. When we spoke in 2009, he told me he'd wake up every day to chat with his friends back home using Yahoo! or Instant Messenger and had been doing so for about 10 years. "I think since 1999 going, the Internet got big in Ghana and then the young folks go online and chat too. It started getting to me. I send a lot of text messages too. We were saving money."

In this chapter I examine the experiences of the diaspora's digitization via an analysis of interviews with my research participants, like Mantse and others at various diaspora sites in the US, the Netherlands, Ghana, and online. This chapter's insights also emerge from participation in virtual spaces for Ghanaians on the Internet, via websites such as GhanaWeb, MyJoyOnline, Twitter/X, Facebook, blogs, and more. As described by my research participants, the effects of these connection strategies have been multiple, impacting life in the diaspora, trajectories of settlement abroad, relationships with tech users in the homeland, and a calibration of identity with regard to the homeland. Here I expand my critique of a purely object- and science-driven notion of *techne* from the book's introduction to the realm of virtual communities: digital diaspora as a technique of online sociality. As a virtual space, the phenomenon includes, of course, online diasporic communities and actors assembled into virtual ethnic- or nation-centric social networks. But working from Black Atlantic literature and thinking around notions of diaspora, the phenomenon can also be conceived of as a *medium of sentiments*, deploying a set of practices meant to bridge ruptures between historic and physical communities.[4] More expansively, *digital diaspora is a technology itself*, one that I later argue more succinctly achieves high impact as a medium for the exchange of affect, which is required to make transnational communities and relations cohesive, tangible, and real. As described earlier in this work, robust network sociality is not a given or preordained in Ghanaian cyberculture, since the main global engines for this online networking are designed and optimized for the Digital North. The Internet has not been well configured in Africa, with a few outstanding exceptions, such as the island of Mauritius (see chapter 5). The idealization of the network society as a "horizontal" or flattened infrastructure, rather than being ascendant from the technological infrastructure itself, is reconfigured by Ghanaians deploying a range of techniques to effect translocal ties, often drawing on the discourse of Pan-Africanism. Digital diaspora is central to these techniques.

Diaspora has been a central concept to describe the transnational migrations from Ghana, certainly in the last 30 years, notwithstanding the *longue durée,* in which the Gold Coast of West Africa was indelibly tied to the international trade in enslaved people. While these forms of diaspora are not the same, they have increasingly become equated in recent years, as the diaspora concept has come to be used more widely to describe the experience of second and third generations of Africans living abroad, especially in the West.[5] The concept of diaspora became integral for Black scholars beginning in the 20th century especially, drawing on the long history of international formations that has served to frame a Pan-African political consciousness, driven in many ways by the descendants of the enslaved, in places such as Jamaica and the West Indies, the Caribbean broadly, and North America.[6] The work of Caribbean educator turned African statesman Edward Blyden, the political activism of W. E. B. DuBois in the late 1800s and early 20th century, and the research and scholarship of midcentury linguist Lorenzo Dow Turner represent a range of actors who harnessed the concept of "diaspora" as a notion of *global African unity* up through the 1950s, when it became more widely deployed. Blyden drew upon biblical inspiration, describing "Ethiopia stretching out her hands,"[7] a phrase that acknowledges how Africa had been "a source and nourisher" of global historical forces and identity for Afrodescendants. DuBois's activism, and the publication of his *The World and Africa,* Turner's "Africanisms in the Gullah Dialect," and the work of Africanist anthropologist Melville Herskovits,[8] further established the notion of diaspora as central to African American cultural theorizing. In chapter 4 I demonstrate how the more recent social embrace of Black diasporans as "Afrodescendants" and participants in shared experiences of dispossession has been accelerated by the solidarity of social media–based movements such as #BlackLivesMatters and #FeesMustFall.

The discourse around the African diaspora has been transformed in the decades since the 1950s, especially with the publication of Paul Gilroy's landmark cultural studies survey of transnational arts and literature, *The Black Atlantic*.[9] In this important text, Gilroy argues for a broader, more hybrid construction of collective Afrodescendant identities (as opposed to "nationalisms"), which the term *diaspora*'s polysemous usage often allows. The word has grown from a term specific to the academy and community organizing to an organic buzzword that maps to a broad range of experiences in late capitalist globalization. Diasporic experiences among Jewish people, archipelago nations such as Greece, and other forms of migration retain an attachment to the term through a long heritage of dispersal, origin, alienation, and longing, that is, narratives of homeland and hostlands. The African diaspora is not minor among these, given the legacy of the more than 12 million enslaved Africans forced to settle in the Americas between the 16th and 19th centuries. As Gilroy has illustrated, the nation-focused, even "Zionist" character of diaspora rhetoric in the Black world has endured since at least the 18th century, as Blacks sought to reestablish forms of political autonomy.[10] There is little in Gilroy's book about Africa, though it remained a central node in the triangle of modernity he discusses.

The questions about what diaspora constitutes would be raised in work on "new African diasporas" by volume editors such as Okpewho and colleagues, Olaniyan and Sweet, and influential writings by historian Paul Zeleza.[11] Still, the expansion of the diaspora concept benefited tremendously from the work of Brent Hayes Edwards (discussed later in this chapter) and from M. Jacqui Alexander's Black feminist positioning,[12] and it has been troubled and deepened more recently by scholars such as Pinto and Figueora-Vasquez, working in the "still-colonized" Latin or Hispanophone Atlantic.[13]

ON DIASPORA

Historically tied to the movement of Jewish people since antiquity, diaspora as a term and concept was applied more broadly in the 20th century with the growth of transnational and postcolonial displacements in the post–World War II era that were voluntary, involuntary, of necessity, contingent, and strategic.[14] Elements of the concept's theorization remain debated across disciplines, with ongoing arguments about to what degree dispersion needs to have been "forced," as in the Transatlantic Slave Trade or Armenian genocide, or driven by "labor migrations," as is the case for the Indian diaspora;[15] or even highly "imagined," as in the case of work on virtual world migrations and revisionist religious movements.[16] Another question at the core of diaspora discourse is whether or not diasporas are ambivalent to or nostalgic for homelands; questions persist also over whether identity maintenance or "collectivity" takes place at all. As transnationalism scholar Laura Candidatu states, "Understandings of diasporas have changed over time, from classical essentialist, comparative to poststructuralist to circulating and multi-spatially situated."[17] This partially encompasses the approach in which African American scholars began to systematize the term, via the efforts of the historian Joseph Harris, Africanist scholar George Shepperson, and others.[18] The contention here is that identities, in this moment of globalization, are increasingly deterritorialized and constructed/deconstructed in flexible ways—ideas reflected in Gilroy's influential work. Africanist sociologist Robin Cohen states that diaspora is a complex "network of relations" and should be acknowledged not merely as material and political, but also as having multivalent ties that nonetheless promote group cohesion. The idea of origin remains central to these sentiments: "[H]ome and the stronger inflection of homeland remain powerful discourses and ones which, if anything, have been more strongly asserted."[19]

The field of African diaspora studies has provided a robust literature since the publication of Joseph Harris's groundbreaking *Global Dimensions of the African Diaspora*,[20] both historicizing the concept in its usage among Black reformers and arguing for its wider applicability in academia more broadly. As George Shepperson noted in Harris's volume, diaspora in the civil rights era was used as a means of linking descendants of enslaved Africans to the continent of Africa, against the intellectual erasures of these ties by White supremacist historians. Shepperson highlights the desire for a transnational discourse on Black experience, which he terms diaspora; linking eras of Black transnationalism to the concept, he states that the contemporary approach to diaspora and those since Blyden's time were interested in "the establishment and investigation of the areas to which the dispersed went and in which their descendants are living; the study of the interaction between these centers and their peripheries, at all possible levels; and the unceasing attempt to integrate these studies into the overall history of humanity."[21] Coming out of the struggles against racial oppression and colonial rule, these early scholarly descriptions of the "Black diaspora" attempted to acknowledge the originary history of what Robert Farris Thompson also termed "the Black Atlantic," in an African as well as enslaved experience.[22] From this period of academic scholarship, diaspora research began to incorporate the experience of not only "New World" Blacks, but transnational experiences in Africa, historicizing the colonization, emigrationist, Ethiopianist, and Garveyite movements; Black communism; and Pan-Africanism as tropes of diasporic movements from and upon the continent. As the post–World War II global order emerged, new migrations of Africans into the West, prompted by the World Banks and IMF's Structural Adjustment Programs (SAP) and political turmoil at home, highlighted the role of African postcolonial states in the social construction of diasporic consciousness.[23]

Diaspora as a collective identity remains fluid and multivalent: contemporary polities such as the African Union (AU) are harnessing diaspora, not only as a trope of culture, identity, and politics (including calls for reparations), but also as a development tool. With a focus on investments, the AU's language is ambiguous in its delineation of diaspora but extends a promise of partnership and seeking aid: "The African Diaspora are peoples of African descent and heritage living outside the continent, irrespective of their citizenship, and who remain committed to contribute to the development of the continent and the building of the African Union."[24] The rhetoric is Pan-Africanist in focus, yet the policies in most countries revolve around enabling contemporary migrants to return home or support their homelands through financial remittances.

The discourse of diaspora has an interesting history and life in Ghana. While the concept of diaspora had been core to articulations of transnational Black identities via organs such as the Institute of African Studies (University of Ghana), and certainly via scholarship since the 1980s, official public discourse in Ghana did not utilize diaspora as a term broadly until fairly recently. By the 1990s diaspora as a concept became part of popular Ghanaian discourse, as Bayo Holsey documents in her work on the creation of heritage tourism.[25] In 1991 PanaFest, the annual Black arts celebration, was initiated by the Ghanaian playwright Efua Sutherland in the city of Cape Coast, home to a UNESCO world heritage slave fort built in the 16th century. The government of populist autocrat turned democratically elected president Jerry Rawlings soon thereafter embraced diaspora tourism as a form of affinity building as well as financial gain, following his visit to Jamaica to attend an annual Emancipation Day celebration.[26] This growth in "heritage tourism" built on the work of African American, Black British, and Afro-Caribbean expatriates living in the region since the 1960s. It has contributed to a contemporary sense of Ghana as a Black homeland and Pan-Africanist nation, as well as a potential tourism destination for race-weary Afrodescendants living abroad.[27] Since 2012 the Ghanaian government has used organs such as GhanaDiaspora.com and DiasporaAffairs.gov.gh across successive administrations to strengthen investments and monetary exchange from Afrodescendants and Ghanaians abroad,[28] culminating in the 2019 Year of Return tourism campaign. Diaspora remains attached to historical sentiments rather than strictly familial relationships and kinship. It is seen as a politically conscious and historical term, part of the legacy of African dispossession via colonialism.

Africanist social scientists working in the diaspora, in particular, have attempted to address productive, regressive, and contingent experiences of diaspora by framing it as "a process that generates subjects through negotiations arising from particular structural and historical conditions that change over time."[29] Using engagements with diasporic subjects historically and via in situ social research, Deborah Thomas and Kamari Clarke have provided ample evidence of ways that diaspora marks the "an identity of passions" while acknowledging that "these may not be identical within particular communities."[30] Such research illustrates the diversity of how contemporary African diaspora communities comprise multiple age cohorts, with diverse sociocultural, political, and demographic differences; sometimes these reproduce divides along ethnicity, gender, age, religion, and politics that are generated in the homeland. At other times these divisions are less salient. These heterogeneous experiences challenge the necessary conditions for "high solidary" for diasporic identities, as formerly argued by Robin Cohen and other theorists of diaspora.[31] This sense of heterogeneity in Ghana's diaspora is echoed both in my observations of communities in diaspora and the digital connection strategies they utilized in order to maintain links to other Ghanaians worldwide. Diaspora thus can be thought of as "heterogenous networks," in ways advanced by both African diaspora theorists such as Tina Campt and by STS theorists of technology such as Bruno Latour and John Law.[32]

DIGITAL DIASPORA

While the extent is disputed, the notion of *digital diaspora* as both a practice and a community of ICT use has been theorized as a means to allow social actors—once structurally bound to a geography of dispersion, exile, and alienation—the capacity for real-time and multisited sociality. Digital diasporas as a form of networked sociality or virtual community may provide an opportunity to overcome historic divides in time, space, and social cohesion that beleaguer dispersed groups. From my ethnographic encounters, it is clear that the diaspora's digital techniques have energized relationships with homeland tech users and helped to calibrate the construction of identity with regard to the homeland.[33] As discussed in earlier chapters, the colonial and northern bias of the Internet's infrastructure configures a priori diaspora-homeland communications, especially in the Global South. From an STS perspective, the notion of *configuration* denotes a soft-determinist interpretation of the impact of technology.[34] *Calibration*, on the other hand, demonstrates the degree to which users selectively adopt, innovate, and redeploy material tools and social media to produce transnational connections in the digital age. In short, despite the structures of dominance, Ghanaian tech users abroad are able to calibrate a way through this hegemonic system in order to effect new networks, especially diasporic relations.

But the calibrations are reassembling Ghanaians in the homeland as well, who have begun to reframe their own self-conception and the social imaginary of the nation through this third space, the Ghanaian and broader African mediascape. Ghanaians and their various migrants living abroad (*aboukyire* in Akan-Twi) have historically participated in global flows that extended the social imaginary of the nation-state and its global ties via trade, slavery, education, political refugees, and labor migrants—especially those referred to as "burgers" or "been-tos."[35] The 21st-century embrace of Pan-African sensibilities around race and nationalism by Ghanaians at home and abroad has undoubtedly been the result of reimagining made possible by these same circuits.

Throughout this chapter I seek to advance the idea that deterritorialized media and practices such as digital diaspora construct what Manuel Castells describes as a "space of flows," but also that those networks are assembled in ad hoc, tactical, and asymmetrical ways, rather than evenly or horizontally distributed, as they are often idealized by cyber-utopians and corporate marketers. Digital diaspora phenomena have been described as "e-diasporas," "net-diasporas," "web-diasporas," "virtual diaspora," and a "networked nation,"[36] among several other constructions. These works variously explore sociality, transnational connections, and the linking of media and public spheres centered around the practice of information sharing. With the emergence of the Internet in the late 1980s and early 1990s, social network formations via digital media were first described as *virtual communities* by the writer and early web theorist Harold Rheingold.[37] Across electronic bulletin boards, Usenet groups, and MUDs,[38] precocious early explorers of these new digital networks learned to socialize and develop community through the asynchronous message systems and relay-chat forums of the early Internet. The austerity of this iteration of cyberspace fostered easy notions of these spaces as "virtual" and disembodied, populated as they were mostly by text-based graphics and email programs. According to Rheingold, *virtual community* often describes online groups who share affinity through common interests different in ways from groups drawn together by kin and location. Early adopters of the Internet in the emerging global African mediascape were also innovators on these systems via message boards such as Okyeame.Net, NaijaNet, LeoneNet, Somalinet, and Dehai,[39] and they similarly built systems from the ground up, often from their elite posts at Western universities. But access to these for the everyday migrant involved less a system of proto-email and more a process of bricolage or polymediation.[40] Toward the end of this chapter I provide a thick description of a streaming event in which these experiences and notions of digital diaspora ultimately are deployed on the ground and in cyberspace, during the Virtual Ghana Fest in 2020.

BLACK DIGITAL DIASPORAS: SOME ORIGINS

Theorists of African diaspora have often described the phenomenon as a longing sentiment or "identity of passions," to borrow the language of solidarity from writer Ralph Ellison.[41] In my research on digital dance cultures, this manifests in both transmission and reception of highly affective media, in which embodied notions of being are indexed.[42] Orienting these exchanges through notions of affect may also be helpful in exploring what was perhaps the earliest iteration of the term *digital diaspora*, in the work of Black British artists and academic provocateurs, such as, among others, Keith Piper, Roshini Kempadoo, and Kodwo Eshun. The concept of digital diaspora emerged within these contexts, not initially as a scholarly description of group sociality, but as an artistic movement, centered in England. Artist and scholar Janice Cheddie explains that the name was used for a collective run by Piper and Kempadoo in the early 1990s, documented in the 1993 online essay "Notes of the Development and Use of a Digital Diaspora."[43] She states: "For Piper, the advent of digital technology gives rise to 'an opportunity' for these globally dispersed and disparate communities 'to re-connect with each other in a pan-global communications network, sharing and exchanging experiences and information'—the digital diaspora."[44]

Working with fax machines, copiers, CD-ROMs, early digital photo editing tools, and mixed-medium sculptures, the Digital Diaspora collective first iterated a transnational Black consciousness with ICT to mark their treatment as foreigners (not citizens) in Margaret Thatcher's England in the 1980s. Such works as "Locating the Remains"—a CD-ROM project and map of African diaspora transits—advanced a defiant, Afrofuturist aesthetic that saw the promise of postcolonial redress in the coming era of electronic networking. For these "UK Blak" activists, the formulation of digital diaspora intersected squarely with Gilroy's notion of diaspora as a particular space of Black expression and tactic of communication.[45] Owing to its expressive function and democratic agency, diaspora remains a practice against essentialism, yet a powerful force for the mediation for affect, whether construed as artistic expression or as having synesthetic impacts on the bodies and minds of those separated by space and time. For Digital Diaspora collective member Kwadwo Eshun, digital artwork emerging out of the Black UK experience offers a set of tactics to combat the "real conditions of existential homelessness, alienation, dislocation, and dehumanization" associated with the history of slavery, African colonialization, and forced migration.[46]

Within anthropological work in Black diasporas, early new media theorists Daniel Miller and Don Slater provided a materialist consideration of the digital diasporic connections, centering both on-the-ground tactics and the content of communication itself. While not termed "digital diaspora," the use of IT networking tools for transnational communities in the late 1990s allowed for an analysis of diaspora formations that are imagined, virtual, and materially bound, focused on the on-the-ground lives of migrants from the Caribbean nation Trinidad and Tobago. In *The Internet: An Ethnographic Approach*,[47] Miller and Slater explored what it meant to be "Trini," via the transnational social formations of Black labor migrant communities in New York and Florida. Key to this was a range of connection technologies, including telephones, fax machines, video and audio cassettes, recordings, and early email listservs, used by Trinidadians living abroad from the islands.

Like the actors in Trinidad and the Philippines,[48] the new media practices or calibrations of Ghanaians I interviewed in this book are multiple and at times multivalent. Each diasporic actor whom I interviewed and interacted with in person and online described the need to utilize specific techniques to interact with particular individuals and with particular communities of Ghanaians. For example, a research participant from Chicago might use Skype for contacts in London and Instant Messenger for friends in Accra. In general, Ghanaians in diasporic locations described connection strategies to reach users in the homeland that were not always similar to the digital media used to connect with various parts of the diaspora. Indeed, later I describe how trending topics for users in Amsterdam and Accra differed significantly between Ghanaian communities in New York and Chicago at key moments on Twitter/X. Other tactics included distinct SMS practices such as "pinging," social media use, blogs, listservs, online radio, video performances on You-Tube, posts on message forums, the use of particular mobile apps such as WhatsApp, and even web-based services representing the homeland. Sites such as Amsterdam's GhanaWeb "country portal" or the music site GhanaMotion.com, formerly based in New York, served as important sites of connection for both diasporas.

Yet what is also clear from my interview data and participant observation is the role of *place* as central to the user experience of Ghana's cyberspace (see chapter 1). As the media scholar Anna Everett has discussed, the notion of diaspora as a racial formation and as a transnational community tied to technology innovation allows us to understand the notion of digital diaspora as a relevant set of discourses with relation to Ghana and Africana studies in general.[49] The Internet's early cyber-utopians, such as Stuart Brand and Pierre Lévy, championed the *deterritorialized* possibilities of information technology during the 1990s—that is, the Internet as context-free, disembodied exchanges, available to the connected regardless of one's geography.[50] These pronouncements about the transnationality of the web were echoed by Castells's *The Rise of the Network Society*,[51] and most importantly his notion about the Internet's inherent *horizontality*—that is, its ability to elevate competing voices to a level playing field of public discourse through distributed media networks. Since then, notions of a symmetrical or "flattened" space of opportunity on the Internet have been challenged at the level of political economy and race, culture, and gender, as well as in important emerging discourse about the role of technical infrastructure.[52] In these discussions, the role of place and material resources has increasingly been asserted as central to understanding and interpreting the social imaginary of the Internet.

In the following sections I attempt to describe several processes of digital innovation among Ghanaians living abroad in diaspora in order to connect with their compatriots in the homeland. The interplay between the tactics of Ghana's homeland digital actors and diaspora actors has contributed to a unique cyberculture, whose circulation is local, global, diasporic, transdiasporic, transnational, national, and homeland oriented. Rather than describe this network in terms of a global polity ("Global Ghana"),[53] I describe the ways Ghana's distinct cyberculture is configured by the technical infrastructure and calibrated by the digital habitus of the homeland, its diaspora *and* its digital diaspora. In Victoria Bernal's important ethnography of Eritrean cyberculture in the late 1990s and 2000s, the political anthropologist describes digital diasporas not simply as a medium of diasporas in the 21st century, but also as a sociotechnical extension of the Eritrean public itself, as the East African nation negotiated its national identity in the digital age. She states that "the internet is allowing for the creation of an elastic political space that can serve to extend as well as to expose the limits of territorial sovereignty. . . . [D]igital media have given rise to the nation as network."[54] This is a powerful framework for understanding the ways nation-centric national discourse extends beyond the physical boarders and jurisdiction of the homeland. But as will become clear in the sections below, the social imaginary indexed by Ghana's digital diaspora in particular references and draws upon the sense of diaspora as a global Black social formation. The valences of Ghana's digital diaspora extend beyond identities already problematically constructed by colonial borders; Ghana's digital actors and online communities gesture toward the collectivist political sentiments central to Pan-African futurism, via technomodernist and social projects aimed at achieving greater African political autonomy in our contemporary moment.

GHANA'S DIASPORA IN CONTEXT

Ghanaians living abroad represent a substantial public with respect to the homeland. Current estimates are that there are more than a million Ghanaians living outside the country and at least two generations of Ghanaian migrant families living abroad in sites in Africa, Europe, and North America. Often underdiscussed in these figures is that the largest population of Ghanaians living abroad resides in Nigeria, followed by the United States and then Europe. Ghana's Bureau of Diaspora Affairs estimates that there are between 1 and 3 million Ghanaians living in the diaspora, which may represent as much as 3–4 percent of the homeland population of 34 million people.[55] US Census Bureau data from 2021 put the number of Ghanaians in residence at 180,700, though the number seems to be an undercount, likely due to undocumented migration or unreported nationality. (For instance, 109,000 people labeled their

nationality as "African" that same year.)[56] Community leaders at Ghanaian events in Northern California and the Bay Area estimated there were
over 3,000 Ghanaians in the region in the mid-2000s; a small population that interacted with a larger African immigrant community including Nigerians, Ethiopians, Senegalese, Congolese, and others. Relevant
to this study, there are an estimated 30,000 Ghanaians living in the Chicago metro area, according to my interviews with the Ghana National
Council of Chicago, and a similar number of Ghanaians living in the
Netherlands, chiefly in Amsterdam and Utrecht combined.[57]

Migration throughout Ghana's history, however, did not begin with
the arrival of Europeans and colonialism. In the centuries preceding the
arrival of Portuguese trading vessels in the late 1400s, local polities such
as the Asante, Dagomba, Ewe, and Ga nations, among others, all migrated from the Sahel region and regions east of the present-day nation-
state. Each ethnic group has a distinct origin narrative that typically
describes a movement from beyond the current borders of the Ghanaian
republic, as well as internal migrations in the centuries since European
contact. Historically, the region known as the Gold Coast was one of the
busiest ports in sub-Saharan Africa and was a central site of the Transatlantic Slave Trade, which produced fundamental diasporic formations
starting in the 16th century. Ethnic groups from the region now known
as Ghana reestablished communities throughout the "New World," most
notably in the colonies of Jamaica and elsewhere in the Caribbean.[58] In
the homeland today, some Ga clans of contemporary Accra include descendants of Yoruba speakers who return migrated from Brazil and the
Caribbean in the mid-1800s.[59] In the late 19th century, as the British Empire sought to establish European political dominion for the facilitation
of colonial trade in the region, Ghanaian elites within the chieftaincy
system and those educated in mission schools routinely began to travel
abroad to be educated in Europe.[60] In this period and in times since,
traveling abroad has become a sign of achievement, as West Africans
began to populate educational systems in Europe and the Americas.[61]
Kwame Nkrumah, Ghana's first president and independence leader, obtained both a bachelor's and a doctoral degree in the United States and
later lectured in the United Kingdom. Ironically, this privileged circulation only exposed Nkrumah and his predecessors to the inequities of
the colonial and world-capitalist system, as his employment was proscribed within the strictures of the racist Jim Crow apparatus.[62] He attended a historically Black college, Lincoln University, upon the advice
of Nnamdi Azikiwe, Nigeria's first president. As a student and laborer,

Nkrumah attended rallies of Marcus Garvey's UNIA movement, whose philosophy of Black nationalism included entrepreneurship and industrial enterprise, the most famous effort being the establishment of the Black Star shipping line.[63] As Ghana struggled to achieve self-sufficiency and industrialization in the postindependence era with Nkrumah as its leader, "modernization" through industrial development became a key goal, with a solid focus on Ghana's infrastructure, and the same was true for other African states.[64] Post-Nkrumah, Ghana's mechanization was thwarted by coups, famine, and political repression.[65] This era accelerated out-migration of Ghana's skilled professionals (the "brain drain") as the SAPs began to produce significant infrastructural challenges in sub-Saharan Africa. Massive state decentralization and divestment in state projects, coupled with changes in US immigration policy, generated a critical mass of elite, educated, and nonelite workers in the 1980s and 1990s, as the West reorganized its labor economy around cheap, foreign labor from the Global South. African migration was in many ways aided by legislation such as the US government's Diversity Immigration Program Act of 1990.[66] The medical industry, for instance, has been a key source of attraction of Ghanaian labor, with typically half of Ghanaian educated doctors and nurses traveling abroad upon finishing their schooling, primarily to meet the demands of the aging baby boom generation in the US and the United Kingdom and the increasing stratification of the medical services industry.[67] However, Ghanaians and other West African immigrants in the US have been steadily active in the prestige industries of finance, engineering, and information technology, and especially successful in the American educational system. Ghana's digital diaspora has partly been constructed from the ingenuity of diaspora-based engineers and innovators, some of whom were instrumental in the creation of the earliest online communities for Ghanaians via websites like GhanaWeb, started in the 1990s in Holland; listservs for the alumni of Achimota Academy in Accra; and the closed web community Okyeame.net, initially built as text-heavy bulletin-board discussion groups at Harvard and MIT.[68]

But the participation of masses in the production of Ghana's digital diaspora should not simply be examined from the content production of these lead Internet users. The contemporary lifeworlds in Ghana's cybercultural practices have been constructed via the widespread adoption of ICT among a range of users in diaspora, many of whom were not involved in tech enterprise in Silicon Valley or elsewhere. The social web, or Web 2.0 in the early 2000s, saw a massive adoption of programs, apps, and websites for online sociality including blogger.com, MySpace, hi5, Facebook, and eventually WhatsApp. This explosion of "user-generated" content further emboldened the rhetoric that cyber-utopianists championed in the early emergence of the World Wide Web as a democratizing force.[69] Yet it is clear that while these forms of digital connection across distance have been influential at some point in time for the majority of my contacts, contemporary diasporic consciousness has been most significantly transformed by the mass adoption of mobile telephony in Ghana during the early 2000s, a period when Africa as a whole led mass adoption of mobile phone use.[70]

The adoption of ICT for transnational identity and state formation represents a technological and social innovation by typically disempowered global agents. As media studies scholar Karim H. Karim states in an early examination of social formations in diasporic virtual communities, "Diasporic media have frequently been at the leading edge of technology adoption due to the particular challenges they have in reaching their audiences."[71] The diasporic communities I have interacted with in San Francisco and Chicago operate as nodes in a global network of the Ghanaian social imaginary. The Ghanaian immigrant communities in localities such as the New York tristate area, Washington, DC, and Toronto, Canada, have older histories as diaspora sites, with distinct ethnic corridors, hometown associations, churches, restaurants, clothing stores, and other shops;[72] contemporary diaspora communities in these towns operate more as posturban immigrant sites, reflecting the logic of neoliberalism and contemporary network flow. Most Ghanaians in the diaspora whom I interviewed interacted with homeland-focused digital media. Chicago, which hosts the largest annual Ghanaian American summer festival, is often described by my participants as having a traditional hometown and ethnically affiliated civic culture. Yet sharing space in an urban landscape so fundamentally shaped by structural racism, Ghanaian businesses and communities have tended to colocate in African American enclaves such as the south suburbs. (A small African corridor in Northside Lincoln Park neighborhood is one exception to this.) Like many contemporary diasporas, the Ghanaian community of Northern California does not occupy a specific "minoritized space" of an ethnic neighborhood. This decentralized experience within an already dispersed diaspora has produced a unique form of ethnic identity in migrant lands. Still, in his 2008 book on the migration of Ghanaians to less cosmopolitan cities such as Cincinnati and Cleveland, Ohio, Yeboah states that immigrants sought to advance their socioeconomic opportunities by avoiding the ethnic enclaves in large East Coast cities such as New York or Philadelphia, moving instead to suburban neighborhoods. He describes a pattern of settlement in the late 1990s and early 2000s in the region that mirrors the migration paths of older waves of more professionally skilled Ghanaian migrants toward areas with solid schools and newer housing. In these sites, he states, "there seems to be a desire for Ghanaians to merge into or integrate, rather than isolate themselves from other Americans."[73]

In 2014 I met Mantse in Sunnyvale for a Ghanaian March 6 Independence Day celebration. The event was organized by a loose organization of Ghanaian professionals under age 40 at a chic hotel. With a hiplife DJ playing for a crowd of about 50 attendees, we talked about the differences between this celebration and a larger event held four years earlier in San Jose, with 200 guests. Against the backdrop of a slideshow showing popular dishes from Ghana, scenes from the crowning of the Asantehene, and the logos of several Ghanaian web firms, he explained that there was another even taking place that same night in Oakland; as he described, the crowd there would likely be younger and more "Pan-African": "We've all got different things going on now. The Ghanaian community here used to be organized around clubs and hometown groups. Now it's professionals and churches."[74]

INFORMATION HABITUS AND THE PRODUCTION OF GHANAIAN IDENTITY

What are the unique ways that Ghanaians use ICT to link across diaspora? No single tactic can be identified. Rather, a constellation of practices and techniques of connection are used to effect diaspora-homeland communication. Based on my interviews and participant observation with Ghanaians in California, Chicago, Amsterdam, and the homeland, tools such as SMS (text messaging), Twitter/X, WhatsApp, Skype, Viber, Facebook, email, blogs, online radio, and various websites are deployed as a repertoire of techniques to connect with ties abroad. Consider the following tactics described by my informants in the Bay Area between the years 2009 and 2014. Sitting in an open-air craft market in San Francisco, Ebo, a former international aid worker, receives calls on his mobile phone from associates in Ghana seeking advice on business start-ups and development projects. Ahmed, a truck driver in the East Bay, looks for the best deals in international phone cards so that he can speak to his children nightly. While doing business in Europe for a global medical firm, Nana Yaw makes a call to Accra, first dialing into a VoIP connection from his home in Berkeley.

Through these diverse material or technical practices, these Ghanaian actors are doing the work of digital diaspora. As anthropologists Deborah Thomas and Kamari Clarke describe it, diaspora is as much community as it is "a process that generates subjects through negotiations arising from particular structural and historical conditions that change over time."[75] Thus digital diaspora itself can also be considered one of these connection technologies—broadly construed as tools, knowledge of tools, and innovative material practices. This is clear from an interview conducted with DJ Mantse and his associates referenced earlier in this book. In our discussions in Oakland in 2009, the group of friends talked about their various strategies of connecting with the itinerant Internet back home:

> *Mantse:* When I used to have a lady [girlfriend] there [in Ghana] I used to call like every day. I call a lot because of my interaction with the hiplife folks, in the hiplife industry.[76] Plus, I have a cousin who handles all my affairs for me. Like if I'm trying to get anything here, like some CDs, get a phone somewhere else, he does it for me. So, I'm in constant communication. I text him a lot. I email him a lot. He comes on the 'net, we chat every day. He updates me.
>
> *Royston:* Would you say you do more phone calls or text?
>
> *Friend 1:* Yeah me, I do phone calls, and I do a lot of text messaging, and I have some other guy, like he said, handling things, so if I want to give somebody some money, I just text this guy, "Give him a hundred dollars for me."
>
> *Mantse:* Exactly. Well, the guy that handles my stuff, I text him and he texts me back. The other people they worried about their credit to text. They rather call you and say, "Hey call." Or they call you and hang up or tell you to call them back. (*Friends say: flashing.*) That's what they call "flash." Have you heard about that term? It's very annoying, they do it at midnight. (*All laugh.*)
>
> *Friend 1:* And half of the time, the name don't come up, it says "Unknown." It's the most annoying thing, I don't know who it is, and they do it like five times. Sometimes they're in New York and they forget. They think we're in New York, so 4 o'clock or 6 o'clock in the morning [in California]. They'd be calling you, because they calculated New York time. And they flash you, thinking you see the number and they hang up.
>
> *Friend 2:* And you *have to* call back.[77]

In our conversations, Ghanaian diasporans in the US described a variety of their calibrations used to connect back home: Some experimented with teletypewriters largely intended for use in deaf communities (TTY), as well as phone cards and third-party SMS websites in the era before smartphones were more accessible on the continent. But these tactical media are all deployed to courier messages between users in the diaspora and homeland via desktop transactions. The central issue in connecting from abroad with those in the homeland is calibrating which tools can be reasonably and reliably agreed upon in order to establish transnational discourse. The tactics of digital diaspora have transformed tremendously since the 1990s, and definitely so since my research began around 2009. In the years since, there has been substantial growth in the penetration of Internet and mobile phones in Ghana, and certainly broader adoption of smartphones in the US and throughout the world. That period also saw the rise of social media's main channels, Twitter/X and the Facebook enterprise (WhatsApp, Instagram, Messenger). This "platformization" of social media, as Wendy Willems describes,[78] has resulted in an emerging monoculture of tech practices, as large companies such as Facebook and Google have come to dominate global market shares. In local markets, however, tools such as Viber and Saya in Ghana, 2go in West Africa, and Mxit in Southern Africa were dominant for nearly a decade. By 2014 the majority of these tools had succumbed to Facebook's WhatsApp and Messenger, or newer channels such as Microsoft's GroupMe and Tencent's WeChat (China). As of this writing, mobile payment platforms and digital wallets have come to dominate most discussion about tech enterprise in Ghana and in Africa in general (see chapter 5). Yet despite technological changes in the last decade, the importance of the sociotechnical system in which digital tools are embedded has remained key for digital diasporic practices. In the capital Accra, Ghana's social elites, tech entrepreneurs, and some businesses and university students I interviewed were more likely to have consistent access to both wired and wireless Internet than working-class laborers or residents in rural or peri-urban regions. But even this is challenged by key sociotechnical issues such as constant power outages or *dumsor*. Issues emerged depending on users' proximity to the electrical grid and roads, often complicated by weak broadband coverage and regular service interruptions by their ISPs. Yet rather than this simply being a problem for those at home in Ghana, these same issues and conditions end up making the diaspora more attenuated to the conditions on the ground in Ghana. MagicJack and other work-arounds are never guaranteed, and these tactics must always evolve with changing information praxis.

The choice of tools, as diasporans describe, reflects personal preferences and also what social role their homeland connections play in their transnational lives. For Nana Yaw, a 50-year-old medical professional living in Berkeley, VoIP is a key technology for his connections back home. Yaw lives in Berkeley's affluent hills area, and he travels to Ghana between business trips to Europe several times a year. There he meets with his family and advises on philanthropic work he does in education. For him, platforms for VoIP were his main mode of contact. He could use these while traveling abroad in Europe, and he had a VoIP connection installed at his mother's home north of Accra. Yet he describes the connection to her as intermittent, troubled by bandwidth issues. Like all other respondents, Ahmed, the truck driver from Oakland, typically received intentionally dropped calls or "flashes" that indicated he was expected to call those living in Ghana back. At the time, he told me his busy work life did not permit him to surf the Internet much, so social media and websites were not part of his repertoire, though he used email to connect to other Ghanaians in Northern California. Eventually he adopted Facebook as his primary means of connecting with friends back in Ghana, but he utilized it less for contacts in the Bay Area during the 2010s.

LINKING ACROSS ASYMMETRIES

In interviews with tech users in diaspora, it was the digital habitus of the *other* participants in the transnational network—that is, the connection strategy of the user in the homeland—that often dictated the tools of connection. Often that tool is a mobile phone, should the infrastructure for mobile phone use be available. This is not always the case, especially in the rural regions of Ghana, where nearly 40–50 percent of Ghanaians reside. Kofi Aidoo, a Berkeley, California, civil engineer, whom I spoke with, said he would call someone in his extended family at least once a week via mobile devices he had purchased for them. He stated that reception in villages outside of Kumasi, where they lived, had only improved recently in the late 2000s, making even regular mobile phone use a possibility. Tech users in metropolitan areas such as Kumasi or Accra could rely on landline-Internet telephony (VoIP, etc.) as well as mobile devices, and less frequently, landline phones. To connect with others in diaspora, however, Aidoo participated in an email-based web community that he and a few others had started in the 1990s. The small community of 50 participants, who dubbed themselves the "Waakye Club," shared jokes, birthday greetings, news from Ghana, and coordinated donations and aid projects for back home. While members lived primarily in the diaspora (in New Jersey, Toronto, London, etc.), a handful lived in Ghana, mostly return migrants, according to Aidoo.

Connections and exchanges across Ghana's digital diaspora, however, were not always from highly connected diasporan to low-tech homeland users. Ama, a graduate student in her thirties living in Amsterdam when I interviewed her in 2013, described her social media use this way: "Facebook is mainly for like pictures and 'liking' things, whereas Twitter is more for journal articles and things like that. And then like another news feed."[79] Her tech use can be categorized in still more succinct ways. Analyzing her Facebook and Twitter/X feeds, it was clear that Ama shared personal information about herself and her family ties, be they in Ghana or elsewhere. Her Twitter/X posts, however, evoked the diaspora experience. She was constantly asking about the state of news developments in the homeland, often wanting her associates' opinions on the headlines. These could be questions about award winners at tech conferences in Accra, responses to President John Mahama's State of the Union address, experiences at polling sites during the 2012 elections, or updates on World Cup matches in 2010. Her Twitter/X community was linked to social media enthusiasts primarily based in the homeland, the cadre of activist-developers discussed in chapter 2 who participate in networking events and tech conferences, many of them students at Ashesi University College. Though operating from a more limited and challenged Internet infrastructure, these are Ghana's digital elites, and their practices represent robust forms of online cultural production, rivaling and surpassing digital actors in diaspora such as Kofi, Nana, and even the social media savvy Mantse, none of whom were regular bloggers or content producers in that time.

Historian Everett Rogers and others have described the familiar bell-curve schema for technological adoption: from *innovators*, to *early adopters*, to *mass market*, to *laggards*, to *nonadoptees*. This theory of "diffusion of innovations" has often characterized notions of tech transfer from the industrial North to the underdeveloped Global South.[80] But several examples in my research with the Ghanaian diaspora demonstrate disjunctive and technological flows that do not easily map to the developmentalist "West to the rest" progression. An interview with John Elmina, a tech worker in his fifties living in South San Francisco,[81] provides an example of this *reverse flow* of digital habitus from the homeland. Though Facebook was well on its way to mass adoption when we talked in 2010, Elmina stated that he did not start using the social networking site until his friends from secondary school in Cape Coast asked him to share photos there. They repeatedly asked him to create an account, which would be more convenient for them to reach him, as opposed to calling or writing email. DJ Mantse in Oakland described a similar scenario. Active on the Internet since the early 2000s, he did not learn about the tool WhatsApp until 2012. Since then, it is the tool that he has almost exclusively used to link back home. He was introduced to the messaging service during a visit to Accra for the Christmas holiday season, where associates were active on the app. Mantse told me he was mocked by friends for living in the US and not using the "latest tools." As an ethnographer, I often heard this same story from other research interviewees based in diaspora, regarding social media services that were popular in Ghana at the time, such as hi5, WeChat, and Viber. WhatsApp, though built by developers in Silicon Valley in 2009 and purchased by Facebook in 2014, lagged in adoption in the US and Europe, even as its use has blossomed globally. This was even more true for "mobile money" services, which have transformed Africa's economies since 2013 (see chapter 5). Rather than technological adoption proceeding in a supposedly linear direction, associated with Western coloniality and technological preeminence, these variegated adoptions interestingly have proceeded from high-tech "global cities," leapfrogging Western users in the US or Europe to become adopted in pockets of the Global South among users where sociotechnical needs find a solution.

As described in chapter 1, using a US mobile device to call a phone number (mobile or landline) in Ghana is typically impeded by the prohibitive pricing policies and the technical differences between cellular networks. Prior to moving to 4G networks,[82] some of the larger American mobile carriers (Sprint, Verizon, and US Cellular) typically used the code division multiple access (CDMA) configuration for their cellular phone systems, while most of the world, including countries in Europe and Africa, operated on the global system for mobile communications (GSM) cellular technology. Ghanaian migrants in the United States prior to the late 2010s relied on the two major American networks that made it easy to make international calls, both of which had sociotechnical disadvantages: AT&T charged exorbitant rates for international calls; T-Mobile had cheaper fees but at the time was a significantly smaller cellular network.[83] This was part of the reason many of my US contacts used calling cards in conjunction with other technologies such as VoIP. Few of my participants in the US had more than one mobile phone, which contrasts with a robust and sometimes ostentatious culture of multiple phones among users in Ghana, who often use both a smartphone and a yam.

Indeed, what was generally a cheap option for tech users in Ghana was often a barrier to access in diaspora: though the overall cost of devices is a fraction of personal income, at 2–3 percent in North America, compared to 7–10 percent in Africa at the time,[84] mobile devices (especially smartphones) are very expensive in the United States and often require lengthy subscription contracts. From my examination of Ghanaian social media usage, particularly up to 2016, on sites such as Twitter/X, Facebook, Instagram, and WhatsApp, as well as blogs, it seemed evident that Ghana-tagged social media was dominated by diaspora users, especially those in the US and the UK, where data pricing for the Internet and mobile phone are and remain cheaper, relative to the pay-per-use ("pay as you go") model in Ghana. Between 2011 and 2016, many online communities operated in the Ghana mediascape, such as the BloggingGhana web ring; even older websites such as Okyeame.net were often mentioned by my research contacts. However, the vast majority of the Ghanaian Internet connections for my diaspora-based research participants were to other members of the diaspora, both in their primary locale and in other parts of the world (e.g., New York, London, Toronto, Amsterdam, etc.), rather than to users in the homeland. Social media in general produced robust connections for especially young adult Ghanaians in the diaspora. Sionne, a 25-year-old college student whom I interviewed in Chicago in 2016, stated that her main means of connecting back home was principally through Facebook. She is active with a diaspora-based church group, which she frequently posts about on Facebook. Her connections to other Ghanaian social organizations via Facebook consist of entertainment companies based in New York and Atlanta. While she stated she frequently listened to the Ghana-based radio station PeaceFM online via a mobile app, more often than not she used a local, Chicago-based Ghanaian Internet radio station, LegendTalk Radio (now defunct), as her chief source of news about the diaspora and Ghana.[85]

Living in the West, these individuals would have ready access to broadband Internet services, allowing them to maximize use of videoconferencing features of YouTube and higher-bandwidth tools such as Skype, Google+Hangouts, etc. Diasporans were conscious of their network privilege given connectivity issues in the homeland, if only to recognize the tenuousness of IT connections in Ghana. In Amsterdam, Ama told me, "The home [landline Internet] is just awful.... It's down, it's up, and then they cut it off. When I went home in the end of 2012, it was terrible. When I went home in 2011, it was great, so I don't know [why.] ... And you would call [customer service] and nobody would be answering the phone or they never come back to you."[86] That Ghana's Internet culture is primarily dominated by content from those in the diaspora is not revelatory given the long-running awareness of a global digital divide, as documented in the work of Alonso and Oiarzabal.[87] This is driven primarily by socioeconomic access and its attendant digital praxis ("pinging/flashing"; multi-SIMing, tethering etc.) Similar research examining differences in online content production and tech use in the US along race and class divides illustrates the ways economic and social elites dominated early Internet discourse.[88] For instance, when the question was posed to diaspora actors about the importance of websites created especially for Ghanaians, there was a typical roll call of diaspora-based domains such as GhanaWeb, ModernGhana, and GhanaMusic.com. Homeland-produced sites such as MyJoyOnline, PeaceFM, GhanaSoccerNet, and OMGghana were also described. Slightly less frequently mentioned were Ghana-specific blogs and content producers, such as Ato Ulzen-Appiah's MIghTy African blog, Circumspecte, and those of others in the Blogging-Ghana online community—that is, digital elites based in the homeland. Appiah's site is significant in that he first started blogging while a student in the US in the early 2000s and has since expanded his blogging to other sites, which he operates from Accra as a returnee.

Among my Twitter/X contacts in the 2010s,[89] the most consistently active were musicians in diaspora, including @Twi_Teacher, @Ghana-MixTapes, and the Chicago/Accra musician M.anifest (@manifestive), as well as the websites for GhanaMusic.com, a New York–based company. Twitter/X users from Europe (UK and Netherlands) were also very cognizant of developments back home and had many connections to bloggers in Accra.

The paradox of digital diaspora for Ghanaians is that while diaspora online praxis has been a means of asserting transnational autonomy from both the politics of the homeland and the social pressures and racism of hostlands (particularly the US and the Netherlands), their interest in Ghana has sustained the importance of place-based identities, even amid ideals of a deterritorialized network society. Even more, Ghanaians at home and in diaspora have largely relied on digital platforms that are not owned by, designed for, or programmed by Ghanaians or other Africans, such as the properties of Meta or Google.

A key example of this is GhanaWeb. Popular digital diaspora content producers such as Adinkra Radio in New York and The Progressive Minds show in Chicago, are hosted on Facebook and YouTube. Like ModernGhana.com, based in Amsterdam, these have provided important services as virtual communities for Ghanaians. But none has had the reach of the Amsterdam-based GhanaWeb. Media Ownership Monitor, an NGO working with the Media Foundation of West Africa, claimed that the news portal was one of the most visited Ghana-centric websites in the country in 2021, competing only with Multimedia Group's My-JoyOnline.[90] When I interviewed GhanaWeb's business director, Roberto Bezzicheri, in 2013, the site was still mostly identified with the diaspora rather than the nation-state. Bezzicheri said at the time that 60 percent of the site's traffic came from the US, followed by the UK and then the Netherlands, followed by Ghana.[91] In the years since then, the site has a been an important source of news in Ghana itself, providing coverage of elections and live events, mostly via YouTube.

In the 1990s the website emerged from a local Amsterdam newspaper focused on Ghanaians living in the Netherlands, started by Robert Bellaart, a Dutch journalist and programmer. Francis Akoto, a writer and engineer working for Nokia, partnered with Bellaart for many years in the transition to GhanaWeb, but by 2002 Bellaart was the site's chief programmer and content officer. As a country "portal," the site is unique in that it hosts a mix of credited and uncited articles, including hundreds of Wikipedia-style entries on Ghana's history, demographics, and culture, as well as user-generated blogs and web-aggregated news stories from outlets and radio stations in Ghana. Its user forum, "Say It Lound," was a source of controversy for decades, as articles and links posted by the message-board users were often received with uncritical boosterism, virulent political partisanship, and ethnic trolling. Even with this mixed history, GhanaWeb was an essential resource for key segments of Ghanaians in several diaspora sites, including Amsterdam. I interviewed Kofi Yeboah, a migrant repairman, who ran a small shop that reconditioned old appliances in the Bijlmer district. (His store was located on Efua Sutherland Street, no less!) In the 2010s Yeboah switched his business to repairing desktop computers and phones, and customers would often request he help them "get on GhanaWeb." For many migrants, Internet access was synonymous with the web portal itself.[92] Part of its success, Bellaart offered then, was that the site was tropically tolerant and designed for low-bandwidth users. He stated that it also utilized the space of diaspora to give Ghanaians freedom of expression that might not otherwise be enjoyed in the homeland:

> The design has been fantastic, because it loads pretty fast, even if the Internet is slow. . . . Funny thing about Okyeame.net, I couldn't be a part of it, you had to be a Ghanaian. You need to have certain level of education to qualify.

The strength of GhanaWeb is that anybody can join, you just don't have to be a professor. If you are a cleaner and you have an opinion to write, you can submit to GhanaWeb, and people will publish it. . . . If it was a Ghana-based organization, we should have been shut down in 1999. We also don't depend on the government in any way. That makes us more independent.[93]

For decades Bellaart was personally invested in the service, having married and had children with a Ghanaian woman and lived in the Zuidoost (Southeast) neighborhood for many years. But he was also a media entrepreneur who extended this approach of a transnational African platform to other properties, such as CamerounWeb, MyNigeria, and TanzaniaWeb.[94] Ultimately the optics of this ownership challenge the notions of Ghanaians utilizing digital media to create autonomous/African spaces for connection. Whether Bellaart could rightly be considered an authentic producer within Africa's and Ghana's mediascape would be for others to decide. Certainly GhanaWeb has remained an important and influential media outlet, illustrating how the material entanglements of digital diaspora may extend beyond ethnicity and gesture toward an "Afropolitan mediascape," as I discuss in the next chapter.

SYNCHRONICITY AND *DÉCALAGE*

One of the key transformations in the diaspora experience for Ghanaians living abroad has been the increasing synchronization between *diaspora time* and *homeland time*. Several theorists have examined these concepts as one of the key affordances of digital networking tools. The Internet and digital media, according to theorists such as Lev Manovich, have been said to "collapse" time from the standpoint of the user.[95] Time is made ephemeral with the archiving of digital content or lack thereof.[96] Laguerre states that this in effect presents a colonizing influence on diaspora time, as working regimes, holidays, and cultural festivals come to reflect more closely homeland chronospheres (e.g., Chinese New Year, Día de los Muertos, Ramadan, West African Homowo, or yam festivals).[97]

For Brent Hayes Edwards, a literary and critical theory scholar of Black internationalism, it is "the practice of diaspora," indeed, via the exchange of some form of media (letters, books, newspapers, journals, etc.), that characterizes the experience of diaspora across time and space. For Edwards, diaspora has always been marked by the time lag between lifeworlds across the Atlantic that was only bridged in the 20th century by newspapers, magazines, letters, memoirs, and musical recordings.

Drawing upon the writings of Négritude writer and independence leader Léopold Senghor, Edwards uses the concept of *décalage* to describe a lag or gap "in time and in space," which Edwards describes as a "structure of unevenness in the African diaspora."[98] This is important in our contemporary discussion, because even today décalage remains distinctive of global diasporic flows, even as Internet practices are often described as being "in real time." This is the condition of disjuncture also referenced in Arjun Appadurai's concepts of global flows.[99] If Edwards's décalage is both the medium and the message, disjuncture is the condition that remains.

Nearly every participant in this research, especially those under age 40, remarked that their chief reason for going online and using ICT to connect with associates in Ghana was to consume news and thus produce a consciousness of contemporary events in the homeland. News information was the object of desire for Ghanaians living abroad and illustrates how vital a role media play in the process of collectivizing sentiments. This was also key to the language used by participants in other African diaspora forums, such as the Dehai's message board documented in Bernal's work. ("*'Dehai adi intai alo'* 'What news is there of our home country?'* was the usual introduction to any conversation between Eritreans").[100] While many Ghanaian diaspora users of online social networks that I interviewed expressed an interest in news or events experienced in the diaspora (GhanaWeb provides a separate "channel" for diaspora-related news), events in the homeland were often described to me as their primary interest. For many, this consumption of news from Ghana seemed to serve as a gauge of their sense of being Ghanaian. Despite the diaspora base of many of Ghana's new media outlets, few were diaspora focused.

This is not to say that the content of Ghanaians in diaspora's use of social media was quite synchronous with developments in the homeland. During a water shortage and fire crisis in urban Accra in January 2014, I was struck by the lack of retweets of posts from the homeland by Ghanaians living abroad. Rather, Ghanaians in diaspora consistently reposted headlines from sites such as Bloomberg.com, such as this: "Ghana orders plant remain open to prevent wider water shortage." Another user in diaspora wrote amid the crisis: "On Twitter I read about there being no water, no electricity in Gh and people here were telling me warmly that Ghana's 2nd only to SA! [South Africa]." In October 2012 I observed similar experiences of disjuncture when I was embedded with activist-developers posting content around an important presidential candidates' debate. While we watched candidates argue over the merits of free secondary-school education, agriculture-based development, and infrastructure on live TV, diaspora Twitter/X feeds were unpreoccupied with events back home. I observed Twitter posts in real time, observing the screens of my research participants by following along with Tweet-deck—a Twitter/X discussion thread–tracking tool—to compare time-lines of other Ghanaian Twitter/X research participants, from both the diaspora and homeland. My research participants and I noted how posts on important issues such as budget proposals, as well as ad hominem attacks, spiked activity among Ghana's digerati in Accra and elsewhere throughout the country. The lists of Ghanaian Twitter/X users from California, Chicago, London, and Amsterdam, however, told a different story: among the UK Ghanaian postings, an Arsenal game dominated timelines, with only scant mention of the presidential debate happening back home.

In New York and New Jersey, Hurricane Sandy had just hit and was dominating Ghanaian Twitter/X feeds from the United States. None of the Twitter/X users in Chicago, California, or the East Coast mentioned the presidential debate at all. Twitter/X users I had identified as being from within Ghana, however, evenly peppered their political commentary on the local debates with posts supporting the English Premier League match. Tweets about the hurricane and its victims in the US were a distant third place on timelines from Twitter/X users in Ghana and Europe. The soccer match by far dominated Twitter/X feeds between Accra, Amsterdam and London, perhaps indicating a close affinity between users in the homeland and in Europe, at least on the issue of football.

Taken as a whole, Ghana's social media users could be said to have been participants in a global online discourse; together with the digital diasporic discourse on Twitter/X, and through other social media sites, the web content could be said to encompass the Internet's imagined community of Ghana. Yet my participatory observations revealed distinct affinities, largely based on their *local geographic concerns*. And the issue of geographic specificity would continue to turn up in other events in the digital life of Ghana during those years: In 2014–15, as an energy crisis emerged in the homeland, the hashtag #dumsor ("lights off") would trend nationally, signifying growing public anger over electricity shortages.[101] But while the hashtag and public demonstrations grew in Ghana, the #StopDumsor movement, including scheduled protests and the maneuvering of politicians around it, failed to resonate on diasporic social media and was mostly absent in US Ghanaian Twitter/X feeds. Rather than uniting the various communities of a now global Ghanaian mediascape, geography and place in these instances overdetermined "national" interests. In these examples, the décalage between diaspora and homeland life, even with the benefit of real-time Twitter/X, remained ensconced. A quip from Ama, made during 2020's national elections campaign, seems as relevant to 2013, when we first met, as it does today. She posted online in December 2020: "Twitter is the worst place to gauge how #ghanaelections2020 is going to pan out. Aside from the echo chambers, I am sure some people pontificating about the elections (including myself) are not even registered to vote."

Diasporic life is built up as an experience of exile. The impact of alienation is negotiated within the construction of meaning in diaspora, which for many often emphasizes long-term acculturation to the political economy and mores of the host society. In the US this includes acceptance of the process of racialization, that is, anti-Black racism, and also resisting it with color-blind doctrines.[102] But the technologies that produce digital diaspora also further reflect the fragmentation of the immigrant experience.

Growing up in Kumasi, Kofi rarely spoke to his parents via telephone while attending a boarding school hours from his home. For Mantse, letter writing was always a key part of his connection to Ghanaians at home and abroad. As a child in school, his teachers taught him basic writing via pen-pal letters, and the practice continued for him when he came to the States as a young adult. For others, audiocassette tapes "pause-mixed" with updates from their relatives in Ghana or newspaper readings—sometimes produced over the days and weeks—were also a vital part of the repertoire of techniques to maintain a connection with the homeland.[103]

In many ways, digital diaspora interactions strengthen immigrant networks, producing a strong diaspora sensibility that reaffirms social identity boundaries; for my research participants, these were specifically present around consciousness of their West African and Pan-African identities. Kofi's participation with various Ghanaian community groups in Northern California and the Waakye Club provides ready forums for cultural practices as well as opportunities for expatriate bonding. They are information networks that allow Kofi to immerse himself and his family in the contemporary lifeworld of the homeland. In diaspora, online messaging and participation at important life events such as naming ceremonies, marriage introductions, and funeral dedications were facilitated by Skype and YouTube. Facebook is used as a public family testimonial. One of the earliest distinctions between my social use of Facebook and that of my interviewees was the very public listing and connecting of family members, with cousins, uncles, and in-laws all listed on the former "top friends" feature and in the personal "About" section. For marriages especially, Facebook photo albums have been robust sites for exchange between transnationals who once would have made a major investment in traveling abroad. Those who couldn't afford travel now have access to key life events that strengthen family ties. Depending on the level of incorporation of technology in the event, the formerly underprivileged in diaspora may now possess a position of prominence unafforded to guests at especially large events on ground. Thus, participants in my work often described themselves as being "in-between," occupying some "third space" between diaspora and homeland, between homeland and hostland. At times the individuals described this as a "cosmopolitan" space or an "other" category, responding to prompts about their identity as if they were checking off items on a census. Yet the space of the "other" or cosmopolitan is *articulated via a diasporic frame*, as I discuss in the next chapter. Kofi told me, "When I am here [in the US], I tend to focus on things here. . . . America has a way of changing you."[104]

In this final section, however, I provide a thick description of a moment when the synchronicities *and* disjunctures between homeland and digital diaspora became simultaneously lived and practiced. Such was the case in the virtualization of Chicago's annual GhanaFest, an event bringing together more than 20 diaspora organizations for a fair, stage performances, and vendors in Chicago's Washington Park on the South Side.

A VIRTUAL GHANAFEST

In late spring 2020 GhanaFest, the largest Ghana homecoming festival in the US, was canceled due to ongoing COVID-19 pandemic shutdowns. The organizers of the event, the Ghana National Council of Chicago, opted instead for a live, web-streamed affair, hosting celebrities, musicians, and cultural presentations. This transnational event would be broadcast across Facebook, Instagram, YouTube, and Diaspora News Television, and would feature musical showcases and speakers in Chicago, as well as live feeds from Accra, Takoradi, and other locales in the homeland. Necessarily, Virtual GhanaFest was hardly the full-bodied experience of years past. I attended my first GhanaFest in 2008, stumbling upon Ghana in Chicago while attending a professional journalists' conference that summer. A listing on Chicago's free weekly newspaper

FIGURE 8. Members of an ethnic "hometown" association parade through the *durbar*, or fairgrounds, at GhanaFest in July 2015. For more than 30 years, this annual homecoming festival has attracted as many as 10,000 people from across the US to Washington Park on Chicago's South Side. Photo by author.

brought me to an all-night concert. I tried to stay up late for perfor-mances by hiplife rappers Reggie Rockstone and VIP but couldn't make it. At the time this GhanaFest afterparty was held in a large mid-20th-century concert theater, stripped down to its concrete essence on Chica-go's West Side. Vendors sold women's jewelry with Adinkra symbols and small bolts of fabric; another vendor had a display of expensive Stacy Adams "Gator" shoes. Watching the performances, what struck me there and at events since then was the complete synchronicity of the crowds. Attendees chanted verbatim lyrics to every song and moved in unison, rather than grooving about independently. As I was to learn, these were Pan-African affairs, with Nigerians, Liberians, and other African nation-als at the celebration.

Since that time, I've visited GhanaFest in Chicago as a means of ex-ploring the diaspora community's notions of identity and practical creation of connections back to the homeland. It was a site for the multi-sited nation itself. In Chicago, the Ghanaian community numbers at least 30,000, a midsize African immigrant community in the Midwest. When it started in the late 1980s, GhanaFest began as a joint festival for the Ga-Dangme traditional Homowo celebration and the Odwira festival marked by Akans.[105] At its height in the 1990s, GhanaFest hosted over 10,000 people. Today, American politicians, Ghanaian diplomats, and members of parliament make regular visits to the affair. Throughout the years, a beloved local storyteller has begun the event with a drumming solo performance, blowing on horns and telling familiar folktales. At an event in 2016, former Ghanaian presidential candidate Papa Kwesi Nduom greeted friends and clients at a pavilion set up by his transnational firm GN Bank. In 2018 the event was so large that it spun off separate networking and cultural affairs for the Ewe community, an ethnic group from the eastern border of Ghana, also predominant in Ghana's neighboring country Togo. Ghanaian high life music, jazz, hiplife and rap, gospel, and Afrobeats artists have all performed. It has always been a Pan-African event, with attendance by the US African American community, though its influence has been diminished somewhat by a comparable transnational Bantu Festival and various Caribbean festivals around the same time of year. After the summer of 2015, it was increasingly difficult for me to make connections with my research contacts at GhanaFest. In years prior, whereas GhanaFest had either been an obligation as a junior family member or a dalliance in the teen and undergraduate years, the 20–30 age group predominantly skipped the official ceremonies for private gatherings, external events, or afterparties that would roll late into the night, spilling across multiple venues both within the city and in adjoining suburbs such as Naperville and Bolingbrook.

Virtual GhanaFest in 2020 was more of an online performance than a family reunion, reflecting back on my data gathering now. Working with Sheriff Issaka, a Ghanaian student at UW-Madison, and Charlotte Von De Bur, an American graduate student, my research assistants and I covered the event remotely, conferring with each other on WhatsApp, taking screengrabs of the live streams, interacting with participants online, and writing up field notes. In my own virtual participant observation, I deployed no fewer than five screens to follow the flow of events, on Facebook, Instagram, YouTube, Twitter/X, and Ghana-based content producer Diaspora Television Network. In addition to the actual footage of the event on YouTube (more than four hours), we screenshot more than 100 photos and interacted with over a dozen research participants. On Facebook alone there were ultimately more than 800 viewers of the event and 278 comments on its message board. (A year later, the video had been viewed more than 8,000 times.)[106] Our intention was to capture the event in all its liveness, recording phenomena as we would have at an actual physical event. As in other scenarios, our experience was no longer simply virtual—by the end of the day we were physically exhausted from attending to so many information sources, in much the same way we might have been taxed trying to immerse ourselves in an on-the-ground field site, interviewing participants, and taking field notes.

As the official broadcast feed started on July 25, 2020, the screen's image transformed from an NTSC color bar to a prerecorded collage of attendees dressed in Daishikis and carrying red, black, green, and gold colored flags walking toward the ticket stands for GhanaFest in years past. In the footage, some women in traditional wraps and dresses enter the South Side park. One carries a basketball under her arm, and the shot fades to an image of a woman in a headwrap and beads, blowing a welcoming kiss to the camera, followed by a montage of Ghanaian and Asante cultural symbols: the wooden staff of an *okyeame*; a royal umbrella covered in Adinkra symbols and porcupines; a *durbar* or assembly of elders, seated in a U-shape formation around the processional square; and elderly Black men and women wearing golden crowns, necklaces, and rings of office, draped in voluminous Kente cloth. The accompanying soundtrack to the video feed is a glistening orchestration of synth piano and organ, with a full-throated man singing, "You are welcome home," a literal translation of the popular Ghanaian word *akwaaba*, a Twi term used to greet foreigners and returnees in the motherland. A voice croons over the images, "You've been kept down for much too long / Stand up please and say I am free / Don't forget you are welcome / Welcome home."

The image shifts as folk music plays in the background, then turns to highlife music. A camera pans over vendor stands displaying *kenkey* and smoked fish; northern woven raffia baskets; and leather purses and sandals, beads, soaps, and fashionable clothes hanging on racks. A shot lingers at a vinyl poster from GN Bank: "Your Bank. Your Community. Your Dreams, Come True." A man's T-shirt reads: "Export Ghana, Export More." The slogan of the Ghana National Council of Chicago (GNCC) rises on the screen: "Beyond the Year of Return: Building Our Community," a reference to the Ghanaian government's diaspora tourism campaign of 2019.

As the live feed from Chicago starts, a GNCC official stands in front of a stage, captured in side-profile view, and begins to speak to an assembled crowd of 50 or so people at a discreet nightclub in Chicago: "Welcome to a virtual screening of our GhanaFest. We've had to do this because of the coronavirus. But we are glad to do so and we are live from Chicago. Today we have in stock for you a show not only from Chicago, but we'll be going all [the way] across Ghana 4,000 miles from here, to bring you a cross-section of what Ghana's culture and exhibition is all about."

As virtual events go, the festival broke with the familiar pandemic-era setup of talking heads assembled in a collection of video-chat feeds. The participants from Chicago and two locations in Ghana (Accra and Takoradi) stood on bandstands, holding mics with live music and audio. The signature sound of multiple audio signal feeds echoed for a few moments. The local Chicago emcee, Nana Marfo, host of the web show *Omama TV*,[107] acknowledged guests by name, facing outward from the stage, not necessarily toward the cameras. Performers and speakers have dressed for a stage performance, with garb meant to be visible. (Charlotte remarks that the attire was fancy, as opposed to casual chic, which dominated Zoom calls in the previous months.) Participants wore well-appointed Kente and tailored African prints. RalphRita production company, the local Chicago firm producing the feed, used multiple camera feeds between a forward-facing crowd address, stage-left and stage-right, and band close-ups. Occasionally the camera panned across the invited special guests at the nightclub standing against brick walls and cocktail tables, PPE masks providing some coverage of faces and chins.

The research team was taken aback by the live vibrancy of the event, though after repeated viewings, it's clear the Accra portion of the show featuring radio celebrity Giovanni Caleb and the Ghanaian gospel singer Nacee were prerecorded. (Accra time is six hours ahead of Chicago.) The speakers and performers routinely addressed and worked the live crowd, while at times speaking directly to the audience at home. Watching via YouTube and Facebook, the audience presumed the producers were actively monitoring the message boards. (Direct queries were seldom answered.) The enduring pandemic prompted a few cautions from people online:

I love it but please ensure that you are following the CDC guidelines. . . . [T]he singer's [sic] are too close. Please observe social distancing.

Please social distance, please. You are too close together. God bless.

Nana. Please put up your mask.[108]

The events of the day included a cooking class in the style of the African recipe trend that was popular on YouTube for many years, and

FIGURE 9. Gospel singer and pop artist Nana "Nacee" Osei performs from Accra during the Virtual GhanaFest in July 2020. During the COVID-19 pandemic, the GNCC opted for a live-streaming event, which featured performances and speeches from the homeland and diaspora. It was broadcast via Facebook, YouTube, and the Diaspora TV Network. *Source:* GNCC/Diaspora Network Television, "32nd GhanaFest Virtual Concert 2020 Ghana Production (Nacee, Ssue and MoBeatz Live Performance)," July 30, 2020, YouTube, https://www.youtube.com/watch?v=FMI3Ah8oA-s.

a GNCC official and fashion impresario premiered her line of couture in a prerecorded runway stroll. A commercial by ArtBanc, a Ghanaian commercial illustrator, offered portraits for "Weddings, Graduations, Funerals, and Special Occasions." Perhaps most notable was an advertisement for Mobiledoctors.net, a start-up that promised premium at-home health care by MDs and nurses for those living in the homeland. The 1990s Ghanaian rap pioneer Okyeame Kwame served as its brand ambassador.

The Virtual GhanaFest lasted four hours and included a gospel performance from Ghana, with a seven-piece dance band, performing on a local broadcast TV stage, again, shot in multiple angles with close-ups of performers and screen transitions between cameras and live-mixed audio. A well-known gospel singer, Nacee, took to the stage in Ghana, bouncing contemplatively, slowly grooving with the music's loping pulse. He spoke and sang in a combination of English and Akan-Twi, sermonizing throughout a 20-minute set, stating at one point: "I had a dream recently that the entire world was in sadness, everyone was crying about corona, everyone was saying 'Corona!' The world was crying. Lovers were losing lovers. This had brought a massive problem, but Jesus's blood has spoken. It has spoken for you and me. Hear me, corona will be subdued. . . . GhanaFest, we have God. Therefore, we do not fear anything. What a mighty God. We will follow you always. . . . If you don't already have corona, you will never have it. All those who have it, will be healed."[109]

From watching online and later talking with research participants in the Chicago area, it was clear that some of the diaspora families had organized watch parties, and in our observations, people were connecting with each other in the comments section. Posts on Facebook and YouTube included "Watching u live from Bolingbrook" and "Watching from the Windy City Chicago. God bless Ghana and Chicago, stay safe." "Watching from Accra," remarked another poster on Facebook. "Watching from GH and I am proud of you guys."

Notably absent from the festival was the procession of traditional elders, the Nananom, or leaders of the various ethnic and civic associations, in the royal attire and costumery. As GNCC President Paa Kwesi Sam stated, filming any gathering of the royals was decided against for health precautions. The absence was notable and lent the event more of a concert feel rather than a recreation of a traditional GhanaFest.

At the end of the event, I participated in a separate Zoom-based dance party, with DJ Bill Bonsu broadcasting from his home in suburban Chicago. Using two Zoom video feeds, Bonsu shot a profile of himself spinning music from a laptop and turntables and broadcast the sound through a separate audio interface. If the YouTube comments evoked a degree of public address (*new orality*) in the presumption of reception and interactivity, the simultaneity of the Zoom dance party was undeniable. More than 80 participants listened to Bonsu mix and blend African and American R&B, hiplife songs, Afrobeats, azonto tracks, and other popular music. More than a dozen attendees used their camera phones and laptops to join the feed, their webcams showing them mostly listening to the music. Mothers and daughters captured selfie footage, while others danced in front of their computer screens. One person had the video on in the background while they worked an outdoor grill. Other participants simply broadcast their faces absorbing it all or sitting on decks in a backyard, smiling in awe.

A PAN-AFRICAN COMMUNITY OF AFFECT

Perhaps the most moving part of the production featured Ghanaian youth and college students providing testimonials to the importance of GhanaFest in their lives. In a prerecorded montage, teenagers sitting at home and young adults in college dorms provided first-person monologues describing their experiences of the festival in years past. Images of Cape Coast beach ran in the background, while fade-ins of panels of the émigré and first-generation youth were shown center screen. These were somewhat scripted, awkward, and stilted confession videos, but hardly insincere statements of love and pride. "This community gives us the chance as young Ghanaians to meet people who look like us, come from the same places as us, and can relate to us. This community also gives us the chance to expand our horizons and find people who can help us get places in life. . . . It's an amazing opportunity for people to understand the ways of living in Ghana and to continue this tradition as we get older," said a slightly awkward teenage boy. A college student sitting in his dorm stated: "For me, having parents come from Ghana, being Ghanaian-American and growing up in Chicago, I believe I was fortunate enough to have been exposed to certain traditions and certain cultural values that let me know that I come from a very, very, very special people. And, when you know you come from something great, it motivates you to wanna be greater." A young woman organizer for the festival shared: "Just growing up learning about different traditions and Ghanaian culture, actually going to GhanaFest, partaking in the activities, the festivities, dancing with my fellow Ewe people, the *borborbor* dance. . . . It was just a great experience. You know, each and every year I go and just take it in. And it's actually like we are actually in Ghana for that day. It's a great feeling."

Had these testimonials been on a stage in Washington Park in Chicago, they might have made little splash. In a year of hopelessness, dread, and death, the youth testimonials were tender, honest, and affirmative messages about the power of diasporic identity, at a time which saw the breakdown and failure of both the homeland's and the hostland's ability to affirm safety and belonging. At Virtual GhanaFest, not all of Ghana was represented, but those in attendance and viewing demonstrated the power of technology to link across distance and the ability of media tools to be completely subsumed within cultural media. GhanaFest can no longer be just a live, physically sited event. Its production and consumption will likely always include the digital, but 2021's version was a small, in-person, "Ghana Picnic" with a virtual participation component. The full festival has since returned to Washington Park. Through the glitches, dropped-out audio, and stop-and-start feeds, a sense of décalage was reaffirmed within a digital copresence. The gaps affirm its authenticity as well: a community-driven production, not driven by slick product marketing or political brand messaging. The video, now timeless but also susceptible to the ephemerality of the web, provides an artifact that attests to the distance and flow between homeland and hostland.

If the event and the phenomenon discussed demonstrate anything in particular, it is that not only have individuals in diaspora assembled a repertoire of tools to effect dialogues of belonging, but affective acts, performances, and responses are key for energizing its sense of community across homeland and diaspora. Media serve as a central element of the Ghanaian community's ongoing practices of digital affective exchange. In a 2017 chapter, I described affective exchange as the "lingering impact of external stimuli that attempt to appeal to an emotion and the senses."[110] During Virtural GhanaFest, comments were seldom meta-discursive in their address. Instead the event's participants interacted and reacted to the live music, discussion, and feedback on screen, demanding responses from the performers. On YouTube, someone cajoled the speakers: "Turn up the volume of the mic." Another person, on Facebook, congratulated the organizers: "Ghana is always the first to make things happen."

In conclusion, it is clear that Ghanaians in diaspora deploy a range of media techniques to connect with home and to other Ghanaians living abroad. Most surprising from my interviews and observations were the ways users continually attempted to calibrate their connection strategy, iterating through diverse media ecologies, never settling on one technique for too long as they attempted to establish symmetry within their networks. If the network society as marketed to them was unable to create a "seamless web" for Ghanaian transitional identities, users generated their own tactical media to do so. While GhanaWeb and other digital message boards and forums have long been key forms of interaction for Ghanaians living in diaspora—dating back to 1994 for GhanaHomePage and 1992 for Okyeame.net—contemporary forms of connection such as Zoom, WhatsApp, Instagram, and TikTok increasingly dominate their discourse, precisely because of the ability to communicate affect. While few of the media are African in origin, the iterative and tactical approaches to connection—pinging, multi-SIMing, SMS work-arounds, MagicJack/VoIP, and WhatsApp—and the strategic use of tools such as

Google+Hangouts, Facebook, and now "mobile money" represent local innovations aimed at lessening the experience of décalage.

One could explain Mantse's material circulations of electronics, for instance, as the activities of a rational actor making good on his transnational privilege. But it would be a mistake to reduce these kinds of activities to a set of communication tactics simply borne out by cost and opportunism. As an actor in this network, Mantse mediates affect between the homeland and Ghanaian communities living abroad, just as he negotiates tangible goods. From the UK Blak arts scene of the 1990s, to transnational message boards, and even to the live, Virtual Ghana-Fest, the stimuli produced through the practices of digital diaspora have endeavored to produce affective exchanges—centered around the social imaginary of Ghana. As I explore even further in the next chapter, this notion of Ghana is conceived then as a multisited Pan-African space on the continent and beyond.

CHAPTER 4

Afropolitan Mediascapes

When his father phoned him from Ghana in 2015, Clifford Owusu felt he had achieved a measure of some success. "You are on TV, son. They're showing one of your videos!" his dad enthusiastically told him from Accra. The video in question was an Instagram clip of Owusu (@Kappacinco) and some fellow comedians acting out a fictional scene at an American embassy. *US Visa Interview Gone Wrong* had already garnered hundreds of thousands of views by then and today remains among one of Owusu's most-viewed videos, still gaining traffic at 1.4 million clicks on YouTube alone. In the clip, a blazer and bespeckled Owusu acts as an immigration official, interviewing a Kente-clad, Krobo bead–adorned visa applicant, played by fellow actor Yaw Yeboah. The applicant states in a stunted English accent, "I have traveled the whole places the world [sic]. I have gone to Bompata, Bekwai, Abetifi, Swedru [all small towns in Ghana]. I've even been to Navrongo [near Burkina Faso]. But see, I never traveled to Europe before. That's why I want to go there." Owusu, as the embassy staffer, responds in a puzzled, Bronx, New York, accent, "You do understand that America, is not Europe?" Shocked, Yeboah cries "Oh, Mass'h [sic], they move it to another place?"[1]

While the airing of Owusu's video on broadcast TV in the homeland was unlicensed, it was welcome recognition for the Ghana-born US citizen, who has been living between New York and New Jersey since he was five years old. "I'm not making profit. But I was just happy to know that I was being shown," he told me when I interviewed him in 2017.

When I initially interviewed the dancer and YouTube creator, he had not been back to Ghana in several years. Yet it was hardly Owusu's first success with social media. In 2012, @Kappacinco had achieved Internet notoriety, largely becoming known for videos featuring Ghanaian *azonto*—an early style of Afrobeats music and dance from the 2010s.[2] Clifford's usual act online consists of mostly ironic shorts and slapstick one-offs, with the style of off-camera-facing cut-scenes that have become the staple of social media humor. He told me that most of the scenes were shot from his home at the time—a simple white-walled and beige-marble-floored kitchen with green cupboards in New Jersey. The setting serves as the backdrop for shorts such as *How An iPhone Would Ring If It Were Invented by an African* (3.6 million views) and *The Reason Africans Don't Answer Phone Calls* (1.6 million views). He's continued to remain relevant with "African father" skits (originally appearing on Vine), parodying a strict immigrant parent in a Northern Ghanaian *batakari* shirt. His more successful clips on TikTok make light of actually becoming a dad. A video with his wife and child, *These New Age Kids Songs Hit Different*, currently has more than 2.3 million views.[3] Today, Owusu has over 800,000 subscribers across his various social media platforms and regularly travels to perform as a stand-up comedian, dancer, or emcee for weddings; in student showcases, and in cultural events in locales from California, to Rome, to Nairobi, and more recently to his homeland Ghana. Owusu's unlikely career move from human resources administrator to in-demand party host started with a video of him and friends doing dances in and around suburban New Jersey to the song "Kukere" by Iyanya. "I looked back now and say, 'What are you doing?'" he told me in the video call, embarrassed by the rudimentary moves in 2012.[4] But *Azonto Fever* was a landmark in early Afrobeats viral dance performances and has more than 14 million views at the time of this writing. The video shows up in any web search for the topic "azonto."

Owusu's journey serves as an important departure for what can now be experienced as a global continuum of African media practices that, like the figures discussed later, are "Afropolitan," that is, "global in focus" and "rooted in Africa," to use author and raconteur Taye Selasi's famous description of the term. Selasi formally introduced the concept in a 2005 online manifesto titled "Bye-Bye Barbar" [*sic*]" writing: "We are Afropolitans: not citizens, but Africans of the world. . . . What distinguishes this lot [Afropolitans] and its like (in the West and at home) is a willingness to complicate Africa—namely, to engage with, critique, and celebrate the parts of Africa that mean most to them. Perhaps what most typifies the Afropolitan consciousness is the refusal to oversimplify; the effort to understand what is ailing in Africa alongside the desire to honor what is wonderful, unique."[5]

In the analysis in this chapter, the Afropolitan adage is flipped, with the media sources themselves "focused" on Africa, though the base of their production may be located anywhere, particularly in diaspora. More than digital diasporas, such online cultural productions are not simply reflective of the virtual communities of Ghanaians and Africans that utilize ICT to develop ties across spaces once characterized as "dispersed," noted for their separation and longing; these media represent a broader space for African cosmopolitan digital culture that extends beyond audiences on the continent. As media scholar Wisdom Tettey has described, the cross-border connections between media outlets in Africa and its diasporas utilize the Internet as a *bridgespace*, connecting home to abroad, a process that generates "solidarities as well as contestations about various forms of political articulation, mobilization, and participation across the deterritorialized spaces that these Africans occupy."[6] As such, these practices can extend beyond the valence of national identity/borders and thus be construed as forms of Pan-African futurism, that is, media and cultural productions that enable the agency of African media producers. In the cases discussed later, this describes the transnational digital circuits centered on Ghanaian and African lives. Theorist Arjun Appadurai has described the intersecting institutions, content, and consumption practices of media as not simply constitutive of one kind of public sphere, but rather as a technological space of sociocultural flow. According to Appadurai, this *mediascape* is inherently multivalent, uneven, and particularistic, given the asymmetry of global capital and distribution of technology. He states that in such an information environment, "many audiences throughout the world experience the media themselves as a complicated and interconnected repertoire of print, celluloid, electronic screens and billboards."[7]

In what follows, I use the concept of an *Afropolitan mediascape*—a space of flows for digital media producers and users who are rooted in Africa but are "global in focus."[8] Rather than strictly an identity or cultural movement, Afropolitanism is theorized here as a mediascape in the 21st century from which to engage with what Stuart Hall has termed the "work of representation."[9] The Afropolitan mediascape describes the connections between actors, their communities, technologies, and institutions in Africa's broader media ecology, where audiences exist beyond the bounds of the nation-state and the continent itself. Digital diaspora and hacking development, as discussed earlier in this book, are two strategies that reflect Pan-African futurist's desire to innovate against global digital divides. But these tactics are part of a broader reshaping of African representations via new media. Ghanaian radio stations CitiFM and JoyFM are but two examples within this Afropolitan mediascape: Based in Ghana with several affiliated station brands in local languages, these companies broadcast multiple feeds online and repost morning talk shows to YouTube, making the locally focused productions available to the world, specifically Ghanaians living abroad. At a different scale, media firms such as South Africa's MultiChoice cable network provide television feeds to viewers throughout the continent, including channels and movies in African languages such as Hausa, Twi, and Yoruba. ISPs such as MTN (South Africa) and GLO (Nigeria), with wireless service in multiple African markets including Ghana, configure a digital infrastructure that bridges both fiber-optic and cellular networked publics with access to the Internet, in addition to providing content programming. These same firms also provide content to social media platforms such as TikTok and YouTube, linking their users across the continent, as well as extending their brands to physical spaces through sponsorship and advertising. These Afropolitan firms need not always be based in Africa, as exemplified in the Ghanaian media space with global IT companies such as Vodafone (UK).

Importantly, the Afropolitan mediascape specifically incorporates creatives living abroad like Owusu, who play active and vital roles in the media in their home countries, and other sites on the continent in a dialogue with homeland producers, with significantly less real-time friction than in previous eras of global flow. Afropolitan cultural production such as Afrobeats and Nollywood cinema also emerges from the continent to penetrate the mediascapes of the West, for example on Spotify or Netflix. This reach extends to other mediascapes in China and Asia and increasingly to the Global South as a whole (e.g., in the ways Nollywood films are consumed as forms of *telenovela*).[10] Afropolitan cultural production, platforms, and social organizing are integral to the spirit and praxis of Pan-African futurism, and as such, Ghana and Africa's digital actors move within this mediascape and have been vital to its development. Their presentist and developmentalist concerns are deployed within and coconstruct social narratives in these Afropolitan spaces.

While the concept has been critiqued as an elitist script for Afromodernity, I draw attention in this chapter to its multivalent capacities, reflected in the ways in which Afropolitan actors sometimes elide class-based distinctions in the service of Pan-Africanism, as discussed in my online and physical participant observation with transnational social actors in Ghana and via a critical analysis of media content. The term can also productively be used to describe narratives in digital media that illustrate Africa's global connections from the vantage of nonprivileged actors as well, as discussed later regarding the film project *Sakawa* (2019). This interpretation provides both an instrumental and populist reading of Afropolitanism in contrast to the concept's overidentification with class privilege. This expanding mediascape and its transformative discourse are partially revealed in the controversy over the concept as it emerged in academia and African public culture.

In October 2011 London's V&A museum sought to answer the question, "What does it mean to be Afropolitan?" by cohosting a forum with *Arise*, a young African professionals lifestyle magazine.[11] Soon afterward Radio Netherlands asked attendees at the African Young Professionals gala in the Netherlands if they were "African, Dutch, or Afropolitan?"[12] A fashion show with the same title followed. In 2012 CNN produced a news segment on elite professionals such as Ghanaian British architect David Adjaye about "living the Afropolitan life."[13] Soon afterward, two leading exponents of African transnational life, author Taiye Selasi and MacArthur Foundation "genius" awardee Chimamanda N. Adichie, released novels featuring globe-trotting Afropolitans torn between home and abroad. Yet such images of Africa's cosmopolitan nature were famously rejected by celebrated literary figures such as Binyavanga Wainaina, who in 2012 castigated Afropolitans and their "digital pulp" novels as West-facing, "product-driven" exhortations for Africans who find that "travel is easy" and "people are fluid."[14] As sociologist Anima Adjepong argues in her ethnography of Ghanaian transnationals living between Accra and Houston, "Afropolitan projects are characterized by an expansive politics of inclusion that seeks to position actors as part of a transnational community of Africans of the world."[15] But figuring out who can claim to be cosmopolitan in a way that centers Africa has continued to complicate the concept's wider acceptance. This paradox will become more apparent in the discourse I examine in the following sections. The tension between who or what can or should be characterized as Afropolitan, as if an identity can be foisted upon any individual or community, should necessarily be fraught. As Adjepong states, Afropolitanism "rewrites some narratives about Africa and Africans while consenting to other forms of cultural hegemony."[16]

HELLO BABAR: WHO'S AN AFROPOLITAN?

In Laurent de Brunhoff children's book *The Story of Babar* (1931), an orphaned elephant leaves the jungles of Africa for a large city, only to return triumphantly with the gifts of civilization: an automobile and Western clothing, including a derby hat. A relic of colonial sympathies or regret, Babar the elephant makes a surprise appearance in the Eddie Murphy movie *Coming to America* (1988), a film crafted by the comedian and farcical movie producer John Landis. "Hello Babar," says Murphy, the African prince Akeem, as he pats a real baby elephant passing through the garden of the painfully fictitious kingdom of Zamunda in the film. The reductive ignorance of *Coming to America* is clear; European royal crowns and stiff vernacular English are a curiously flowery depiction of African autonomy in an era that was still attempting to shake off postindependence despots. The film's representations are (ironically) meant to provide a vindicating image of contemporary Africa for a (Black) American audience, while also using the immigrant experience as a background for humor. Babar is also the departure for Selasi in her consideration of the postapartheid circulations of trans-African identities in her manifesto, "Bye Bye Barbar: or What Is an Afropolitan?" which appeared in a blogpost in 2005.[17] Here Selasi commits to writing the sentiments conjured by fellow authors such as Teju Cole (Nigeria), NoViolet Bulawayo (Zambia), and other noteworthy transnationals emerging in the global literary scene at the time. Selasi sketches a lifestyle and ethos for the emerging African transnational actors of the digital era, beginning with a London dance party: "The whole scene speaks of the Cultural Hybrid: Kente cloth worn over low-wasted jeans; 'African Lady' over Ludacris bass lines; London meets Lagos meets Durban meets Dakar. Even the DJ is an ethnic fusion: Nigerian and Romanian; fair, fearless leader; bobbing his head as the crowd reacts to a sample of 'Sweet Mother.'"[18]

Selasi asks critically, "What happened in the years between Prince Akeem and Queen Agbani?" that is, between the release of *Coming to America* and the 2001 crowning of a Miss World from Nigeria, Agbani Darego—the first Black African to win the crown. What follows, in her words, is part profile, part declaration for this emerging public whose generational charge is to complicate Africa's representation, while remaining true to both global and ancestral roots.

Inspired by this neologism, Achille Mbembe expanded on Selasi's concept, extending his quest for "new ways of seeing" Africa in the 21st century.[19] For Mbembe, Afropolitans signal not only those ensconced in Africa, but those who deal with the world on their own terms: "Artists, musicians and composers, writers, poets, painters—workers of the mind who have been aware since the beginning of the post-colonial era."[20] For Mbembe, concepts such as Afropolitanism provide renewed commitments to the continent while offering distance from the notions that have "ossified" political and cultural innovation on the continent, including parochialism, xenophobia, and racial essentialism. "Afropolitanism is not the same as Pan-Africanism or Négritude. Afropolitanism is an aesthetic and a particular poetic of the world. It is a way of being in the world, refusing on principle any form of victim identity—which does not mean that it is not aware of the injustice and violence inflicted on the continent and its people by the law of the world. It is also a political and cultural stance in relation to the nation, to race, and to the issue of difference in general."[21]

For many, Mbembe's and Selasi's credos would be reduced to the slogan, "Global perspective, but strong African roots."[22] But soon after, the term *Afropolitan* morphed, in some circles, from a defiant marker of pride to a source of neocolonial division: at that 2012 African Studies Association-UK conference, Wainaina stated unequivocally, "I am a Pan-Africanist, not an Afropolitan."[23] Novelist Yewande Omotoso, who grew up between Nigeria and South Africa but was born in Barbados, similarly excoriated desires for Western legibility among those hoisting the mantel of the Afropolitan. In a 2014 conversation with the scholar Rebecca Fasselt, Omotoso stated, "Being an Afropolitan to me sounds as if you are supposed to be a mediator between the West and Africa because you have travelled and lived overseas. I have no torn allegiances, and I have no current interest of ever living in America or the UK. I want to live here. I'm of the continent. . . . The term Afropolitan only seems useful for the West as it gives the West an opportunity to understand and even 'consume' Africa. I am not from the West and I don't need anybody translating things for me."[24] While doing the research for this chapter, I briefly met Chimamanda Adichie at a bookstore in San Francisco where she was promoting *Americanah*. I asked her what she thought of the concept, and she first replied, "What is that?" After I explained, she said, "I am not an Afropolitan."[25]

For many, the Afropolitan discourse magnifies an exceptionalist identity, signified perhaps most dramatically by the election of US President Barack Obama, son of a Kenyan émigré, a fact often cited by online commentators who heralded the term. For critics, the move conflates the growing popularity of global African/Black celebrities—such as Trevor Noah, Lupita Nyong'o, and Idris Elba; artists like South African singer Tyla or Davido from Nigeria;[26] football players such as Mario Balotelli; and celebrity designers Ozwald Boateng or Virgil Abloh—with Africa Rising, the important period of growth in GDP and global interest in Africa following the dawn of the new millennium. For detractors, Afropolitan's elitism and successes were woefully equated with economic and political stabilization in Africa itself.[27] And yet the notion persisted as a lifestyle brand and meme in the African digital mediascape I was documenting in the same period. It turned up in advertisements for college and postgraduate parties in cities such as Chicago (The Adinkra Society's #AfropolitanNight), Philadelphia ("Afropolitan Sundays"), and Boston, where a midnight cruised was labeled "An Afropolitan Affair." It was central theme in an "Ethnik Fashion Weekend" in Paris in 2013, whose images included svelte and multiracial fashion models, walking through Europe's "Africa Shops" in Kente-fusion couture. South Africa's *The Afropolitan* online magazine has been among the most vociferous supporters of the concept, using mottos such as "Inhaling Freedom, Celebrating Life." Initially a project of Johannesburg's Kaya FM, the monthly journal and web vertical has been published since 2007.[28] Former managing director Greg Maloka described the outlet's content as catering to "a mature, sophisticated, socially-conscious individual rooted in his or her heritage and a progressive thought leader."[29] Luxury cars and jet skis, executive profiles and designer clothing dominate its images and feeds, alongside stories of families advocating for natural hair styles.

That a transnational identity marker should link aspects of the African social condition is hardly a new phenomenon. The focus of 19th- and 20th-century African and diasporic political movements has been the acknowledgment and exercise of links between Africans dispersed globally and the social circulation and ties created by the flows of modernity: slavery, "Back to Africa" movements, Africa's colonization by Europe and the ethnic fragmentation in its wake, state formation, and Pan-Africanism as a liberation philosophy. All of these generated collectivizing identities for Africans and their dispersed descendants. Gilroy's notion of the Black Atlantic as a spatiotemporal ethnoscape for Black (and non-African) social actors, culture, agency, and trade is but one of these geographies—though his initial critiques failed to incorporate Africa-based actors.[30] Hakim Adi, Imaobong Umoren, and others have illustrated how the Négritude movement and 20th-century Pan-African Congresses highlighted both the unifying and multivalent nature of these connections, especially through media forms such as newspapers, missives, novels, and letters.[31] Added to these circuits were the "Afrocentric" flows of the post–civil rights/independence era—the Swahili-inflected circulations of Kwanzaa; Rastafarianism and reggae;[32] in the "homeboy cosmopolitanism" of the hip-hop era;[33] and various diaspora "heritage tourism" projects,[34] including the 2019 Year of Return campaign in Ghana. There is little doubt that the deployment of Afropolitanism as an idealistic expression and a description of lived African bodies renews traditions of transit—in this case, working against the negative portrayals of Africans in the global imaginary. Yet as Adjepong states, "Those who identify with the term sometimes make claims to being black in particular cultural, historical, and social ways. Not all Afropolitan projects are the same. However, they are all concerned with advancing a modern non-victimized narrative about Africa."[35]

The question of class has vexed Afropolitanism from the outset, with literary scholars such as Yogita Goyal highlighting that Afropolitans in their everyday politics and representative writings "search for a way of living beyond binary options of enduring social, economic, and political collapse at home or striving for immigrant precarity in the first world. The ability to choose between these options is determined by class and pinpoints the limits of a term that describes belonging and dislocation without explicit reference to economic conditions."[36] Yet among some of its earliest adopters, Afropolitanism was conceived in much less elitist and self-serving ways. Rather than simply dilettantism, Selasi admonishes at the end of her 2005 essay, "most Afropolitans could serve Africa better in Africa. . . . A fair number of African professionals are returning; and there is consciousness among the ones who remain, an acute awareness among this brood of too-cool-for-schools that there's work to be done."[37] In 2011 Selasi and Derrick Ashong, a Ghanaian American Internet impresario, further promoted the concept at a symposium hosted by the Houston Museum of African American Culture, titled "Africans in America: The New Beat of Afropolitanism."[38] Ashong, who also rapped under the name DNA and fronted a band called Soulfège, attempted to create an anthem from the concept with the title track to its 2011 album, AFropolitan, singing:

> You running, like you been down-graded
> This is dedicated to all my natives
> Put a little something in your soul to remind you
> where you been, where you at, and what's fated
> . . . We ain't gonna let nobody savage our culture

So rise up, my people, let us see what you got
No one can tell you what to think
Yo! We got you
It's time to build a new foundation. This is dedicated to all my natives
6 billion ways to live choose one
Afropolitan, Afropolitan, Afropolitan.[39]

The triumphalist exhortation of denigrated African identities is a central theme of this track and on the album in general. In this manner, Afropolitanism is positioned not simply as elite privilege but as a means of calibrating a sense of Africanness amid the challenges of the 21st century. At core we see here and in other Afropolitan musical texts the basis for a cross-class solidarity, rooted in populism, construed as pop culture, as I discuss further later.

Ashong cuts an interesting figure in the Afropolitan mediascape. Born in the late 1970s to Ga parents in Accra, he spent his childhood between Brooklyn, Saudi Arabia, and Qatar, where his father worked as a pediatrician for many years. In the early 2000s he, along with Jill Ford, an African American venture capitalist, started a short-lived business NGO in Accra, attempting to fund tech start-ups focused on African markets. Many of the participants were prominent at the first Ghana/diaspora BarCamps between Stanford and Accra in 2009 (see chapter 2). When I interviewed Ashong in 2013, he asserted that Pan-Africanism and Afropolitanism were not necessarily mutually exclusive in his mind, but if the 19th- and 20th-century movement was construed by many as a racially exclusive social formation, then there needed to be a concept that speaks to Africa's "multiplicities." In our interview he described a Lebanese friend who served as president of the African Student Association at Stanford University in the early 2000s and was also an Arab music promoter who grew up in Accra and speaks fluent Ga and Twi:

It would be so comfortable for some people I know to be like, "We're Afropolitan: Only Black folks are invited." And as soon as you say that to some Dominican kid, he's going to be like, "What are you talking about?" As soon as you say that about a White Zimbabwean, he's going to be like, "What are you talking about?" As soon as you say that to [my Arab friend], he's going to be like, what are you talking about, and you're going to sound ridiculous. Because race is only one factor in the formation of an identity. . . . I guarantee you're going to find Ghanaians who are going to relate more to somebody who doesn't look anything like them, who happens to have grown up in Accra, or in Lagos, or in Harare, than they are going to relate to somebody who grew up in the heart of Flatbush [New York], who does not know anything about their life language, culture or perspective.[40]

And yet my own experience with individuals who could be characterized as Afropolitans in Accra revealed ambivalence about the concept. As a researcher visiting Ghana for the first time in 2011, I was
struck by the worldliness of Accra, Cape Coast, and Kumasi, at times
contrasted with cities where I had lived in the US. Ghana was for me immediately deeply Christian and deeply Muslim, multilingual, multiracial,
multiethnic, and multinational. Details leapt out at me: the French air
conditioner, power generator instructions in Cyrillic, the Chinese tourist
and businessmen, the Dutch embassy always used as a landmark for directions, social events at Alliance Française, offices for DeLoitte, cars by
Mercedes and Renault and Ford, Nokia phones, "China phones," Indian
vocational tech schools, Senegalese and Nigerian restaurants, Jamaican
music, American country and western songs, and Latin telenovelas all
played at shops along Ring Road, or even in cybercafés in small towns
such as Kuntanse.

In Accra I often spoke with young people casually talking of visits
to other bustling African cities, what Gikandi terms the "Afropolises"
(Legos, Dakar, Addis Ababa, Nairobi, Luanda)—cosmopolitan cradles
representing a networked public sphere, across postcolonial history.[41]
I encountered members of Ahaspora, a social network and professional
association of African and Afrodescendant returnees in Accra. Started
in 2011, the group's name combines the Twi word for "here" and *diaspora*.[42] As I polled several of my research contacts about the utility of
the Afropolitan moniker, most had not heard of the concept yet. Some
embraced the definition put forward earlier. Others were deeply skepti-
cal, uncertain that this latest nomenclature was a repackaging of African
elitism—or referenced cohorts of the privileged described in South Africa as "black diamonds."[43] In October 2013 I attended an art opening
near Oxford Street and asked Raina, a mixed-media sculptor, what she
thought of the movement. She responded, "That's not what I am interested in. I am trying to build up what's here in Ghana." Minna, a gallery
attendee, quipped dismissively, "Africa is becoming trendy now. When
I was in school in London, if you talked like a Ghanaian, Black people
would tease you for being 'Aff.' They'd say you're so 'Aff,' [as in] 'African.'" Following the show, a group of us retreated to a trendy restaurant.
Although it was just a few blocks away, my host for the evening, a lo-
gistics engineer who had lived in Atlanta for most of her childhood, im-
plored us to drive, so she could keep her Audi SUV close at hand.

Waiting for our meal, we discussed our transnational circulations.
I talked about the Puerto Rican side of my family, while another person

shared her experience of growing up in Canada and her recent trips to Santo Domingo and South Africa. A few, recent transplants to Accra were not Ghanaian at all; they were Nigerians, Ivorians, and Angolans who had spent their formative years in Paris, Toronto, or Washington, DC, and had moved here for work opportunities. As I brought up the notion of Afropolitans' privilege, they demurred with stories of their own experiences with anti-African racism. One vented about being followed in stores in Paris and Morocco, while another talked about being made to stand in "Black lines" at Euro-style clubs in Accra or being skipped in the hair salon as workers catered to Lebanese and "half-cast" Blacks. The discussion was intriguing and also apparently unsettling; a few White/European ex-pats seated near us left quickly as the conversation became more boisterous and forthright. Those at the table also shared their annoyance at Ghanaian diaspora returnees who preferred to socialize with only themselves and foreigners. The implication was that *those were the Afropolitans*, though clearly we also could all have easily fit the descriptions offered earlier. I shared a copy of *Vogue* magazine I had picked up in Amsterdam on my way to Ghana and directed them to an article listing hip things to do in Accra. They were aghast, flipping through images of foreigners touring craft markets and features about restaurants with goat cheese. Seeing goat cheese on our own menu next to cosmopolitans and mango martinis, we attempted to order some. The waiter returned with a sad expression, telling us there was no goat cheese to be found.

Such a concentration of transnational returnees like the women and men described here in places like Accra seems to advance the saliency of Africa Rising for that time, as many had left the West for brighter futures in the homeland. Was this a recalibration of what Marcus Garvey meant by "a shared destiny"? For me, it was clear that *this* African world, and the backlash against the Afropolitan movement, attempted to assert the already embedded experiences of the translocal that has defined parts of Ghana and Africa as a whole since before the contemporary nation-state. Later, I discussed the notion with Ghanaian software pioneer Herman Chinery-Hesse, who had spent many years living in the United States and Europe before starting theSoftTribe in the early 2000s. Hesse was characteristically acerbic about the use of the term: "Africa is Afropolitan and always has been so. Diasporans are not cosmopolitan, they are ready to assimilate, always afraid of being pigeon-holed. What, being Western-educated makes you cosmopolitan?"[44]

AFRICA'S MEDIASCAPE AND THE AFROPOLITAN GAZE

If the description of 21st-century mobile and transnational African social actors remains necessarily fraught as a form of identity politics,
I argue that *Afropolitan* may serve as a more apt term for the African
global digital media circulations in our moment than perhaps just for
people. In 2014 a YouTube scripted show was released to the public
whose protagonists' lives in many ways reflected the worlds of the transnational actors I have described here: professionally accomplished returnees living in Accra with multiple heritages and citizenships. Billed as
"Africa's first web series," *An African City*, according to its promotional
material, "tells the story of five successful women who confide in one another about their love-lives and find new ways to deal with being a 21st-
century woman in Africa."[45] The show mirrors *Sex and the City*, and
the show's main character, Nana Yaa, narrates her transition from life as
a publicist in New York to life in her homeland Ghana, in voice-overs
reminiscent of Sarah Jessica Parker's character in the HBO series. In one
key episode, "He Facebooked Me," Nana Yaa expounds on Africa's contemporary "rise" against the backdrop of chaotic street scenes and fer-
vent construction projects. She narrates: "According to *The Economist*,
Africa is the next frontier in ICT. According to *The Financial Times*,
technology is driving Africa's transformation. According to the publication *African Renewal*, Africans are coupling their already extensive use
of cell phones with a massive interest in social media."[46] As this intro-
ductory monologue suggests, technology is key to the discourse of Africa Rising and Afropolitan circulation. Nicole Amarteifio, *An African
City*'s director and lead producer, left her own career as a publicist at the
World Bank to produce films about Afropolitan subjects. Born in Ghana
and having attended college in the American Northeast, the director was
quoted in the press at the time as stating, "This could be a moment to
change the narrative of the African woman. The African woman does
not always have to be the face of an antipoverty campaign. Rather, she
can be the face of everything beautiful, trendy and modern."[47]

As the show illustrates, the concept of an Afropolitan mediascape indexes the space of flows that Africans are utilizing to redefine their media
worlds and their physical worlds through networked digital media: This
is a discursive space for globalization, as it is "African in roots, global
in focus" and also African in focus, rooted wherever Africans exist (at
home, abroad, and in third spaces, including virtual communities, online

FIGURE 10. A promotional image in 2014 for Nicole Amarteifio's *An African City*, a web-based TV series set in Accra, featuring young professionals returning "home" to take advantage of "Africa Rising." This was in part Amarteifio's story, having spent the majority of her professional life in the West. *Source:* Amarteifio, *An African City*.

platforms, and audiences). As an ensemble of media and digital spaces, it demonstrates a particularity for and from African producers' own perspectives, re-reading the "dark continent" as "bright," representing Africanness (at times Blackness) into especially non-African spaces, and living lives in contradistinction to colonial mapping. As a distinct global network, the Afropolitan mediascape engenders perhaps a sense of commitment and connection to lifeworlds as constructed in the interests of Africa and its diasporas, drawing on the legacy of Pan-African politics, propelled by the optimism of Afro/African futurism, the successes of African enterprises such as M-PESA, and the zeal of activist-developers in advocating for political change across the continent. It is a space of Pan-African futurism, especially as these social projects and actors are active in shaping the continent's social imaginary toward liberatory aims. It would be easy to analyze diverse forms from across the continent (including convergent media such as global television channels, transnational blogs, and online radio) as objects that make information flow "horizontal." However, the multivalent and unequal distribution of these technologies reflects enduring urban-rural, religious, linguistic, political, infrastructural, and economic divides in Africa that produce fragmented experiences rather than seamless webs of connection. We can return to Appadurai's description of the premillennium information ecology of globalization to helps us understand the uneven potential of these multi-valent circuits. "[T]he suffix-scape allows us to point to the fluid, irregular shapes of these landscapes, shapes which characterize international capital as deeply as they do international clothing styles.... What is most important about these mediascapes is that they provide large and complex repertoires of images, narratives and ethnoscapes to viewers throughout the world, in which the world of commodities and the world of news and politics are profoundly mixed."[48]

It follows then that representations in the African digital mediascape would necessarily also be variegated; streams of Akan-language UTV find distinct audiences globally, as do Ga-language films and Islamic podcasts from Ghana's north. As a varied terrain of discourse, Afropolitanism has since the early 2000s carried with it a sense of unearned or privileged entitlement and visibility, which critics have used to delegitimize its vision. Yet in its contestations across popular culture, we see a discourse aimed at populist constructions of a Pan-African identity, even if the cosmopolitan experiences are not necessarily the same. These kinds of media production, which I am describing as Afropolitan in nature, can include work by diaspora producers such as Armetifieo; comedians such as Clifford Owusu; and fellow content producers such as the YouTuber Wode Maya, who is based in Ghana and travels throughout the continent as a lifestyle blogger. Across the continent's mediascape this might also include firms such as iRokoTV, which was variously based in Lagos and London. In the Ghanaian case, Afropolitan media can include the hybrid media (terrestrial and streaming) of JoyFM, consumed by audiences worldwide via online radio and YouTube. It may include Ghanaian diaspora outlets such as the Facebook video channels *The Progressive Minds Show* (Chicago) and Adinkra Radio (NYC). Terrestrial radio to online streamers such as Black Stars Radio on the public access network SALTO in the Netherlands ultimately contribute content to this now ascendant African media ecology. Chicago's *Afrozons* pop music show, for instance, airs locally on FM radio and has been syndicated internationally since 2017. According to the definition put forth by Ashong and others, the moniker can also characterize firms like GhanaWeb, the popular news site and message board for diasporans, which has been a key portal for Ghanaian news, current events, and online sociality since it was founded in Amsterdam in the early 1990s.[49] It was founded by computer programmer and journalist Robert Bellaart, a Dutch citizen who is White and married to a Ghanaian. In 2016 the company expanded its presence in the homeland, hiring an on-the-ground editorial staff in

Accra—formerly, the company had relied on freelance writers and guest columnists. For the first few decades of its existence, the firm consistently reported that most of its traffic originated from the United States or Europe. In 2021 it claimed to be Ghana's top Internet destination.[50]

Digital artifacts and social media posts demonstrate the ways the Internet and technology have been major sources of solidarity and contestation for the concept of Afropolitanism. Selasi's manifesto primarily appeared online; Achille Mbembe's essay "Afropolitanism" appeared in the art volume *Africa Remix*, in which digital pastiche and transnationalism are firmly invoked.[51] Critics such as Paul Carlucci of *Think Africa Press* stated, "The Afropolitan experiment is over" in his review of Adichie's and Selasi's 2013 novels.[52] And yet such bridgespaces served as home for the moniker, including the well-cited *MsAfropolitan* blog, the podcast *This Afropolitan Life*, and a France-based Afropolitans Facebook group, which promoted a nonprofit organization that flourished to 5,000 members, before becoming inactive in 2020. As self-described Afropolitan and playwright Mwenya Kabwe has offered: "[We] are a community of international wanderers who vacillate between feeling at home anywhere and nowhere. . . . . there are in fact ties that bind. Among them, I believe, is simply the commonality of being charged to make sense of a collection of disparate experiences that include a relationship to the African continent. It is a virtual community on a number of levels: we typically do not literally group together *en masse*, and among the first locations to identify in each new neighborhood is invariably the cheapest internet café."[53]

This interpretation of Afropolitan as a description of digital media, technologies, and their circulation presents a challenge to criticisms that the idea simply indexes elite circuits. Afropolitanism here serves as a praxis and a counternarrative, rebranding the continent's global circulations, a project that necessarily serves as a corrective to perennial news stories and ideational positionings of Africa as an enduring site of crisis. Even amid celebratory accounts of "Africa's mobile phone explosion," the continent continues to be characterized as a site of anti-, or at best alt-, modernity. Africa's abject status is continually invoked by images from landfills such as Agbogbloshie in Accra,[54] erroneously described as "the largest e-waste dump in the world."[55] While films such as *Blood Diamond* (2006) have received their due critique as "White Savior flicks" or for rendering African life as perpetually hapless (e.g., *Beasts of No Nation*, 2015), the global discourse of Africa's primitivity has also continued to be advanced on the Internet, with regular controversies such as #Kony2012—a viral social panic in which the trope of the savage African warlord was used to flatten the ongoing multistate conflicts in Uganda and eastern DR Congo.[56] Even video games have cemented these notions, for example, in titles like *Resident Evil 5* (Capcom, 2009) or *Call of Duty: Modern Warfare 3* (Activision, 2011), in which players can first-person shoot mindless hordes of zombies or militia men in settings such as Somalia. These themes of Africa's abjection have persisted throughout the decades on social media with hashtags such as #NotTodayEbola during the 2014 pandemic outbreak and racist Tik-Tok memes accompanying Shakira's "(Waka Waka) This Time for Africa," which reemerged in the last decade, most problematically using blackface parody. They are not restricted to Western media; in February 2020, BBC's Africa Eye web series reported on a robust online market for Chinese content makers, selling videos featuring children from rural Malawi and other African countries. In the clips, children would repeat self-demeaning and racist slogans about their intelligence and promise not to move to China.[57]

While variously neoliberal and Pan-African in character, online filmic interventions such as *An African City* do more than work at the representational level. These productions also provide what bell hooks famously described as an "oppositional gaze."[58] They are filmic tropes, Afropolitan representations, and media counterprograms, utilizing techniques that had been similarly cutting-edge forms of state propaganda during the colonial era. Media scholar Brian Larkin describes Britain's public spectacle of screening imperial news reels during the early and mid-20th century in Nigeria, India, and elsewhere as a means to produce "a particular sort of modern colonial subject. Technologically adept, forward thinking, mutable, this subject was formed by the crisscrossing of new communication networks."[59] Against these tropes, productions such as *An African City* inject an oppositional practice of representation that is now accompanied in the wider African digital ecology by small-scale, user-generated content, social networking sites, and increasingly more accessible tools, such as high-definition (HD) cameras, helping filmmakers to recognize and produce new icons for African modernity.

Technically, *An African City*'s videography is an essential tool in Afropolitan counterdiscourse. Amarteifio's visuals are consistently produced with warm, golden lighting, rendering the show's subjects in great detail, that builds toward emotional richness with a range of close-up shots. This is a vital innovation in the cinema-photographic canon, given the film industry's long history of configuring dark skin as a problem.[60]

Visual fidelity is in part due to what seems the extensive use of lighting techniques attuned to Black skin, typified in the aesthetic approach taken by filmmakers such as Biyi Bandele (director of *Half of a Yellow Sun*, 2013) and Andrew Dosunmu (*Mother of George*, 2013), filmmakers whose work highlights both rich and poor African transnational circuits.

Amarteifio takes care across all 10 episodes of her show to portray Nana Yaa as a naturally beautiful subject, filming her in posh restaurants and in scenes walking across broken pavement, stopping to adjust, then abandoning, high-heeled shoes stuck in a crack in the road. Group mid-shots capture her young friend crew dressed from head to toe in crisp, iridescent dresses and sculpted hair. (Many of the episodes were shot at the African Regent Hotel, whose branding includes the phrase "Simply Afropolitan.") Less-glamorized scenes—for instance, shots of the protagonist's sweat-patched armpits as she navigates local bureaucracy—attempt to humanize Nana Yaa without making her base or transgressive. It is through these digital optics of glamour and the ordinary that Afropolitans celebrate everyday beauty and the mundane. As a returnee-centered drama, the series is striking in its stylization: the kind of African women portrayed are not just full-bodied "strong Black women," but independent, sexually confident, and vocal in attempting to untether themselves from the deep cultural obligations and contemporary social mores of religion, graft, patronage, and gender that are part of the social terrain of urban Accra. The dialogue renders an undeniably African transnational life, with complaints about generators, importing cars (and vibrators), relations with the "house help" (read, servants), trips to sparkling new shopping malls, the shabbiness of the international airport, traffic on one of the many dusty roundabouts, finding a job, going on dates just to get an apartment, abstinence versus indulgence, sex without condoms, and boyfriends with questionable hygiene.

While the lives of *An African City*'s subjects are out of reach for so many, they are also embedded in the everyday (ordinary and casual) consumerism of the working classes (both middle- and lower-income laborers). In these modes, such Afropolitan gestures can also be seen as populist, collective constructions of cultural pride, aspiring to material well-being and recognition, images that are often outside global popular images of Africa emerging from the West and at times, those ascendant on the continent. As well, these Afro-cosmopolitans in their content, presentation, and subject matter reflect a desire to be seen as participating in a national discussion and media about belonging, rather than always foregrounding life abroad. Throughout the series, use of terms such as

"African," "Ghanaian," "Nigerian," and "West African" in many ways stands in for discourse that is diasporic in nature, forgoing discussions of ethnicity ("tribalism"), region, religion, and party politics that are key to shared identities in the homeland.

When *An African City* premiered on YouTube, it was primarily promoted through viral marketing and later through a strong media push. Print and blog reviews appeared in outlets such as *Ebony* magazine and a host of influential Africana-centered sites such as Shadow and Act and Bella Naija. Though it mostly appeared on YouTube, its second season was aired across the continent on the satellite provider DSTV's Ebony-Life channel.

These digital actors are participating in what seems like a national discourse on the surface. However, given the actors' and producers' geographic positioning, modes of exchange, and access via the West, their concerns and discourse can reasonably be critiqued as pertaining instead to the diaspora. In these digital spaces, the global connection and social imaginary of Ghana, Nigeria, and Africa in general can often be routed through diaspora rather than nation-based networks. And yet these visual projects go beyond the discourse of diaspora-homeland media, conjoining the dispersed and grounded notions of a national imaginary that is also collectivized as "Pan-African." Ensconced in the medium of popular culture, such representations signal, perhaps, the ways elite forms of what might be termed *liberal populism* create opportunities for agentive politics through self-representations largely centered in leisure and pleasure, what the theorist Oluwakemi Balougun has termed "embodied respectability."[61]

AFROPOLITANISM, NOT NECESSARILY FROM BELOW

If the aesthetics of *An African City* can be interpreted as populist interventions that serve to move beyond elite pretense, the gesture may not be as dramatic as the representation of Africa's "multiplicities" in the other Afropolitan film projects. In the much less well known 2018 indie-film *Sakawa*, Belgium-born Ghanaian director Ben Asamoah attempts to document the lives of those operating from the underside of techno-modernity. As the title suggests, *sakawa* is the main theme of the film—a practice of Internet scamming, enhanced by metaphysical-cum-material religion. Anthropologist Jenna Burrell states in her work with Internet café users in Ghana during 2000–2010 that "sakawa came to be associated in emerging public understanding with spiritualist practices, the mechanisms of juju and 'blood money.'"[62] Burrell's research contacts at the time denied using sakawa, and instead averred that their "skills of language and persuasion and the evolving format of the scam" determined the level of success of their digital fraud schemes—especially the tactics of "romance" and "advance fee fraud," elsewhere known as 419 or *kindoki*.[63] As digital humanities scholar James Yékú has written, rather than mere criminals, such individuals represent "Afropolitan antiheroes," figures like Adekunle Johnson Omitiran, a Nigerian Canadian "playboy" with multiple passports and a penchant for luxury goods, which he showcased on social media. In contrast to respectable Afrocosmopolitans, scammers' familiarity and facility with digital media and Western ideas about Africa allow them to strategically maneuver for profit within a global digital media, or specifically the Western mediascape. Yékú states that these actors "inhabit multiple (web) localities, [and] need to be considered as African digital subjects whose identities extend, trouble, and complicate the notion of Afropolitanism."[64]

Like the episodes of *An African City*, *Sakawa* aspires to profile the lives of much less successful African digital con artists with authentic clarity. Asamoah and his Dutch/Belgian film crew capture footage in 4K-HD using crisp, natural light and sympathetic color focused on everyday scenes in Ghana. Scenes are shot on dusty red roads and clapboard phone-repair stalls hidden along Accra's side streets. The film's subjects move between the rural and the metropolitan with ease via *trotro*. Traveling shots document preachers in public squares admonishing youth for participating in sakawa, while motorcyclists and taxi drivers dodge sewer ruts, and mothers read to their children by candlelight during *dumsor*. But the film also traffics in the spectacular, in this case, the market, lagoon, and salvage shops surrounding Agbogbloshie, a landfill often documented in media accounts for the toxic practices of e-waste scrap collectors working there. The site, communities around it, and the adjacent, mostly austere dwellings of the sakawa scammers are central features in the film, creating an uncomfortable association between hightech and the global periphery. Theorist Kwame Edwin Otu describes this juxtaposition as a "synecdoche for the violent consequences of neoliberal infrastructural modernity."[65] Having negotiated with fraud-ring leaders to document the participants' lives in such articulate yet mundane ways, Asamoah captures life in medias res. Riding the line between narrative feature, documentary, and cinema verité, no names are given for any of the characters in the film, and none are credited in the end reel. In its promotional material, the Flanders Audiovisual Fund, which supported the development of the film, presented *Sakawa* as a documentary, as did most press accounts during its release. During the Q&A, Asamoah stated that the genre form was unimportant; rather, he intended to capture the authentic emotional character of the low-level "Yahoo boys," who portray themselves, making the line between constructed and documented reality porous in this narrative.

OneDollar, identified as the lead character by the director, seeks the assistance of a mallam and later a rural spiritualist to help him advance his fraud schemes to earn money to travel to Italy. His sister, a young widow who sells onions at a market, wants to become a hairdresser and helps the protagonist by uploading photos of herself and chatting with suitors online. But when she is unconvincing in this endeavor, OneDollar must secure a mobile phone hacked with a digital voice synthesizer so that he can sound more feminine as he sweet talks male marks from the US and elsewhere. In one scene, scam artists ply their trade, sitting side by side in a cinder block room, a ganglia of power strips between them, and a faded Real Madrid banner hanging on the wall behind them. The team of young men and women operate multiple Facebook chats and dating site feeds on 1990s-era CRT monitors and ThinkPads. Drinking beer and smoking, they flatter their online contacts with loving missives and promises of sexual favors, typing across multiple chat windows: "My Dear." "I miss you." "I'm also single with no kids seriously looking for a honest caring and trusted man." To pass the time they pepper each other with questions: "Who is the richest man in Europe?" "Bill Gates." "No. Europe, Europe, not America." Taking a break, they discuss strategies: "I have browsed but I haven't met a Canadian yet," a junior scam artist tells the main character in Pidgin and Twi, subtitled in English. "Canadian's English is not good, because they speak French. But if you are patient, you will get them," he replies. "These whites are very wicked. I am concentrating on US clients for now." "Oh yeah, they pay a lot. Illinois. Denver. Colorado." "I'm searching Chicago." "People in Chicago don't fall for scamming. How far are you with your client from Finland? He wants sex?" "Of Course. He will never come online and ask: How are you?"[66]

If the Afropolitans only signaled privilege and glamour in their initial iterations, *Sakawa*'s depictions cast a fuller and morally defiant image of African subjects, active and interactive with the world at large. Such depictions might be characterized as a kind of "globalization from below," "from the margins," or "from the periphery," but if we are to recognize the populist agency that an Afropolitan mediascape and its digital media allow for, we can rebuke the overtly elitist conceptions of the term, as several cultural producers have attempted to do. Such a "post-Afropolitanism" would recognize, for instance, what Jennifer Wawrzyniak celebrates as a "a global space of intersecting lives and confluent voices."[67] I argue that such depictions also engage in a politics of representation, contending with the ways in which Africa (however racially/ethnically configured) is necessarily always cast as Black and anti-modern. Given the historic underprivileging of "brown-skinned bodies" in this cosmopolitan circuit, Afropolitanism in both *Sakawa* and *An African City* at least speaks to the normalization of Afro-cosmopolitan life, while also demonstrating (perhaps problematically) possibilities for African-centered autonomy.

These tactics may necessarily be ambivalent, as technology allows for various forms of agency and local interpretation. While this book frames Pan-African futurism as a prosocial use of digital media, no serious researcher of African digital media can ignore the degree to which negative and illicit aspects of the Internet also constitute such mediascapes. In this regard, the scammer, content farmer, AI-content moderation gig worker, hacker, sexual and human trafficker, or terrorist represents technological innovation toward different ends. One may also include the numerous gray market practices of Ghanaian and African tech entrepreneurs (including SIM-card spoofing and questionable e-commerce sites) and e-waste dumping and salvaging as technological interventions within the African mediascape. An acknowledgment of multiplicities is key to understanding Afropolitan circuits. Furthermore, the embrace of a racialization of these realities (where they are obvious) and their cloaking via colonial and illiberal practices in the homeland remains fundamental to expanding these optics of contemporary African digital life.

In a Q&A with Vice Media, the director Asamoah stated, "I wanted to get away from the way most documentaries are being shot in Africa. [They're] very much raw, which is the way it is. It is very much rock-n-roll. I wanted to show you Africa, or Ghana in this situation, in a beautiful way. Nicely framed, but still hopefully getting the message across. . . . When they think about Europe, and the UK, when they look at their buildings, their roads, there's a big difference. Do they feel guilty about scamming? I was not interested in that. I just wanted to tell their story."[68] Having spent the majority of his life in Belgium, Asamoah attended boarding school in Ghana between the ages of 11 and 15. On a holiday trip home, he discovered many of his schoolmates were involved in Internet scamming. As he explains, "This is actually my personal reason of wanting to tell this story, because it would have potentially been my own life. If I did not have the chance to study in Europe, I would have probably been doing this practice."[69] As the film's director of photography, Jonathan Wannyn, a Belgian cinematographer, states:

> Almost all the scenes, I don't say all of them, are staged. And then you bring in real people with real stories, but not in, necessarily, in real environments. Sometimes the environment is created by us because it's too dangerous for them to show their activities, like the big room with where they're all behind the computers, that was just an empty room and I hang up light and then the sheets and everything. It's a completely fabricated scene. But of course, what happens in the scene is real. The people have their real customers, they have their real conversations going on.[70]

Sakawa is an alluring experimental project, constructed as an unglamorous but truthful story about technological flows as experienced by some of Ghana's most desperate transnational actors. If the representations featured in the film also suffer the fictions of the Afropolitan gaze, in their desire to document truths, Asamoah and his crew have succeeded in complicating deficit perspectives on tech access and globalization from the Global South. While there is no evidence that the Afropolitan treatises of writers such as Selasi are specifically invoked by the creators or subjects of Asamoah's film, the practices and circulations of these social (and technical) hackers allow us to see other distinct valences of Ghanaian contemporary life as Afropolitan. This film also illustrates the many paradoxes of Afropolitanism: if it was initially touted by transnational elites and cultural producers, its aspirations to economic well-being and transnational mobility remain largely shared across the continent and still serve as a foil to depictions of Africa's otherwise presumably subaltern status. In this broader interpretation, Afropolitanism can be applied to the experiences of African modernity in globalization via digital circuits that are experienced across class, though those experiences with technology, media, mobility, citizenship, and identity are shaped and experienced in different, sometimes competing, ways.

AFROPOLITANISM FOR THE PEOPLE

It may be helpful here to revisit Stuart Hall's notion of the "dialectic of culture" to understand how Afropolitan media can function as both an African appropriation of neoliberal sentiments and its populist, digital counternarrative. Fittingly, the site of contestation is popular culture, rather than simply academia and among the literati. Hall states, " [W]hat is essential to the definition of popular culture is the relations which define 'popular culture' in a continuing tension (relationship, influence, and antagonism) to the dominant culture. . . . Is the novel a 'bourgeois' form? The answer can only be historically provisional: when? which novels? for whom? under what conditions?"[71] In the popular arts, including music, fashion, comedy, soap operas, and so on, and in the circulation of these through radio, television, film, social media, and online content platforms such as WhatsApp, Afropolitans also aspire to communicate agentive and vindicating narratives about Africa. That it has found an easy and leveling vehicle in other forms of media such as music, as discussed in previous sections with the emcee Derrick Ashong, should come as no surprise. In my fieldwork in Ghana, I observed radio stations and television VJs curating music from South Africa, Cote D'Ivoire, Senegal, Kenya, and Liberia, as well as the United States and the United Kingdom in the same programming, increasingly under the label Afrobeats or "Afro-fusion." On the youth-oriented Y FM (Accra), DJ Flexinard at one time produced a midday "Afropolitan mix," a collection of pop and R&B songs from across the continent and Black diaspora. (When I spoke with him, he was totally unfamiliar with Ashong or Selasi or Mbembe.)

As a popular cultural form and a mode of mass media, music necessarily derives from and speaks to its audiences; its aesthetics engender a sense of unity for that audience. This is especially the case when the narrative address is directed toward common markers of cultural identity, whether language, a sound, a locale, a scene, or some other ethnic identity. In Afropolitan appeals in music we also see egalitarian and Pan-African futurist tendencies. The artist Samuel "Blitz" Bazawule is a particularly notable figure in this regard; his Afropolitan gestures have been ensconced in politically charged Pan-Africanism, rather than remaining ambivalent to history and traumas of Black diasporic life. Working as a New York–based hip-hop musician, Blitz the Ambassador, Ghanaian-born Bazawule fully embraced the Afropolitan mantel during the mid-2010s as he toured alongside the self-described "Afropean" vocal group Les Nubians (originally from Paris). Blitz titled his 2014 album *Afropolitan Dreams* (Jakarta Records) but would later turn from music to directing films, most notably Beyoncé's visual tone poem *Black Is King* (2020) and Oprah Winfrey's musical adaptation of *The Color Purple* in 2024. In 2013 Bazawule, still primarily a musician, had viewed the moniker not as a marker of elitism but as a tool for recalibrating African representations. In an interview with the lifestyle channel Afrofusion Lounge, he stated:

It's a new definition of who we are. I feel like anybody has the right, at any point in time to say this who I am, what I am, and own it. I feel like we've gotten to the point as young Africans, even some old Africans, who are thinking a lot more diaspora, rather than just what can I do in my little corner. . . . Our struggles are the same. Our joys are the same, no matter where you find us. So that to me is the important part of what it means to be an Afropolitan. But at the end of the day, [we're] taking it all back home [to Africa], because that's where it begins and where it ends.[72]

Like Ashong's work with the band Soulfège, Bazawule approaches the liberatory dimensions of the Afropolitan concept through images of an eminent Pan-African unity assisted by technology, most notably in the rap song "Africa Is the Future."

Welcome to the future United States of Africa, we salute you
A new state of mind, I introduce you
Close your eyes and visualize the rise of a new paradigm
Solar powered skyscrapers it's bound to blow your mind
Just launched another satellite flight into orbit, the African Space
    Program. It's so ironic
How our currency is worth more than the Dollar or a Euro
Without the foreign aid, unemployment near zero
We never practice capitalism or socialism. We got Africanism, and that is
    how we living
Not Anglophone or Francophone. We got our own
No more corporations exploiting us in our own homes
Time-traveling unraveling the truth, because we figured out
The West cannot live without the wealth we produce
And now our image is controlled by us. So hold steady
Africa's future is now. So get ready.[73]

If the question is asked as Yogita Goyal does, "When was the Afropolitan?," in comparison to earlier postcolonial writing,[74] another question may serve as a riposte: In the 20 years since the concept went viral, why does Afropolitanism continue to resonate so much among cultural producers, despite its derision in some sectors? Rather than being aloof from issues of the homeland, I observed Afropolitan Ghanaians in cyberspace often connected hyperlocally: city assembly politics played out on GhanaWeb message boards; Facebook debates over the rap prowess between Tema and Accra and later Kumasi's "Kumerica"; and music on GhanaMotion.com that circulated from home to abroad and back, with DJs pledging allegiance to US-centric East, West, and Southern hip-hop camps. Still, other practices also seem to signify Afropolitan networks: Mr. and Ms. Ghana beauty pageants in Germany and other host countries; Christian missions and Muslim mosques in New Orleans, Oakland, and other American cities catering to and run by African nationals; and Ghanaian churches simulcasting services between New Jersey and Cape Coast in ways similar to the Virtual GhanaFest (see chapter 3). Such circuits of social media, sport, work, and travel, even political flight, highlight the variegated landscape of African-centered digital media. The diversity of these Afropolitan media forms reveals the distinctions in Appadurai's sense of global "disjuncture" in contrast to an otherwise simple, unitary concept of globalization or the cosmopolitan.

In interviews with Ghanaians in diaspora, themes of being transnational—not simply of being *in diaspora*—repeatedly came up as a new social reality for my interlocutors, as they talked of frequently crossing the lifeworlds at home and abroad. Some spoke of diaspora as a kind of "third space," unwittingly quoting Homi Bhabha.[75] Some expressed a broad sense of being beyond nationalism ("a citizen of the world") in ways similar to Adjepong's interlocutors from Houston (in 2021). Among those I interviewed living in the homeland, in places such as Accra, many likened their multisited "cosmopolitan lives" to those of Pan-Africanists such as Nkrumah or Casely Hayford. Others mentioned contemporary figures such as journalist Afua Hirsch or former UN Secretary Koffi Anan. In the period of my fieldwork from 2013 to 2017, few of my interviewees took up the term *Afropolitan* as a personal label. However, it is clear from their personal histories, lifestyle, and outlook that the experiences described as "Afropolitan" in contemporary African popular culture, film, and intellectual discourse might just as well signify their own circuits. And yet, as James Yékú states, "Digital mobility multiplies the ways in which the Afropolitan can be mobile without leaving."[76]

This can perhaps be illustrated with another nighttime vignette. I am sitting with five young men in their twenties and thirties in an apartment in Tema, the port city just outside Accra. They swap between a mobile phone and other devices (TV, iPad, laptops), sending missives on Twitter/X, updating Facebook, messaging via WhatsApp, and surfing the web. One is a prominent blogger, another runs a coworking space for entrepreneurs, and another is a network engineer with a small firm and several employees. The engineer travels regularly to Nigeria and elsewhere in West Africa seeking business opportunities and working contracts. That evening, I am privy to debates about radio's preference for Accra rappers over Tema musicians like hiplife rapper Sarkodie (a 2012

BET award winner for Best International Act). The tech entrepreneur is busy practicing his DJ skills, while a Nigerian video channel plays in the background. On the screen are hip-hop and R&B videos that all have the semblance of New York or Atlanta but in many ways are worlds away: the music is azonto, hiplife, "GH Rap," soulful music from Nigeria, the UK, and French-speaking Africa—brilliant and sparkling imagery. There is no [insert American pop star du jour], rather, 2Face, Mzbel, and Shattawale. For me, it is a dizzying torrent of unrecognized faces and names, though the aesthetic, style, and stance are utterly familiar to me as a Black American. Looking for more adventure, we take the 40-minute drive into Accra and eventually end up at an expat nightclub. There, a German DJ spins European techno and Rihanna with incipient Afrobeats tracks. My associates opt to leave; this is not their space, they tell me.

This experience signals yet another of the many valences of the Afropolitan mediascape: American swagger "indigenized" by Ghanaian media publics in which Black diasporic forms are also embedded rhetorically and historically.[77] Yet these subjects also speak, sampling, remixing, constructing a new norm that reflects the worldliness of their lives. The cosmopolitan circuits here link New York, Palo Alto, Atlanta, New Jersey, Philadelphia, London, and Tema; these practices consume the global, producing it via a digital discourse on Twitter/X, Facebook, WhatsApp, Viber, WeChat, and other local variants. To be sure, my interlocutors represent, at least, Ghana's rising middle class,[78] and in many ways its digital elites. But Facebook, YouTube, and globally circulated media such as film are hardly the province of the privileged streamer anymore. The media sphere these digital entrepreneurs inhabit has an African-continental focus. These Ghanaian social actors blog about "smart development," politics in the US, Barack Obama, and the NBA, yet their cosmopolitanism is seldom as multiracial or multicultural as the worlds discussed in *Ghana Must Go*, with its Ivy Leaguers, eloping Chinese/Ghanaian lovers, or scenes like Selasi's club in London. The Afropolitan playwright Kabwe, who claimed both Cape Town and Boston at one point, offers the usual troubled question, "Where are you [Afropolitans] from?" To ask this question in Ghana always provokes a translocal response: "I was born in Accra, but I am from Volta Region" (etc., etc.). Who is to say that the Kente-clad visa petitioner in Clifford Owusu's viral video was not similarly translocal? Furthermore, what is the meaning of origins in a nation that considers itself foundationally Pan-African?

FIGURE 11. Women walking past a billboard for the NPP government's proposed Agbogbloshie e-waste recycling plant in 2018. The area is globally infamous for the toxic practices of salvagers, who burn computers and electronics in the landfill to extract precious metals. Photo by author.

AFTER IDENTITY, BROADER NETWORKS

When talking with Owusu in 2017, the term Afropolitan did not come up, despite the comedian's transformation from someone who primarily saw himself as a "New Yorker" to a globe-trotting minor Internet celebrity. In many respects, @Kappacinco typifies the Afropolitan: He maintains a hybrid identity produced through a life lived both in Africa and the West; his videos reflect deep engagement with Africa's digital mediascape as both a consumer and a producer; his creativity is rooted in African culture (historical and contemporary), Western situatedness, and popular performance; and he maintains an abiding desire to showcase positive, in his words, "modern" portrayals of Ghana's culture to anyone with an interest. I asked him then, "Do you really need to be in Ghana to be a part of Ghana's pop culture?" Owusu replied, "No. Because of the power of social media, you don't really have to be in Ghana to touch Ghanaians. . . . But there are certain things, for example if you want to be a Ghanaian actor, for sure, you have to go to Ghana. They won't take you serious being from here. People want the authentic stuff."[79]

Still, that a local TV producer and audience would find Owusu's video entertaining in the homeland means his work has resonated—either for its stereotypes, for its ridiculousness, or perhaps because of its familiarity. Owusu is engaged in an Afropolitan address, which while elite due to his stature and position in the Global North, indicates an idiomatic or vernacular sense of humor shared with audiences at home. The location of humor and critique is polysemous, yet like Ashong and Bazawule, Owusu's influences and interests remain invested in advancing a Pan-Africanist vision of African people as agentive subjects. He said, "I'm one of those people trying to move the culture forward. In America, Jamaicans and Caribbeans have come and made their mark, Hispanics have made their mark, I think it's time for Africans to also make some type of mark."[80]

Afropolitanism represents a mode of being, germane to the current period of globalization as Africans seek to seize the means of representation amid a moment of prosperity and Western intrigue. As Gikandi states, the label indicates that "the withering or delegitimation of the African state has given credence and authority both to the idea of a global Africa and its particular localities."[81] The ruptures of this representational discourse have been especially poignant in the UK, where the Afropessimist missive "T.I.A." has been remixed by transnational musicians such as Fuse ODG into the rallying cry "T.I.N.A.—This is the New Africa."[82] (Potentially burying the Thatcherist ghosts of that same acronym.) Necessarily a form of rebranding, it signals a desire to be *seen* as African and to free African people of the paternalistic images of a continent continually in need of development aid and "good governance." It may also attempt to elide Pan-Africanism as a political identity, especially those articulations of it as a militant, anti-Western discourse, and/ or racially essentialist. More than the celebration of excess (as depicted in reality TV programs such as Netflix's *Young, Famous & African*, 2022), Selasi, Ashong, Bazawule, and others assert an oppositional Afropolitanism that seeks to represent the *worldliness* of contemporary African life, especially the cosmopolitan flows that shape the self, community, and work in the homeland and the diaspora.[83] As cultural critic and hip-hop scholar Msia Kibona Clark describes, "It is the first term, however, that attempts to capture the multicultural African (Black) spaces that have emerged out of increasingly mobile Black populations."[84]

Adichie, for instance, may not claim the Afropolitan, but *Americanah*'s characters' agency with ICT depicts what Castells describes as the digital media's ability to generate a "space of flows."[85] The central character finds success as a blogger in the US. She emails her ex-lover in Nigeria while sitting in an African braid shop in Trenton, New Jersey. The lover receives said message in Lagos via his Blackberry, while sitting in sweltering traffic. Interlocutors in the novel pass news serendipitously during visits from abroad for holidays, or business travel, or funerals. Children in Nigeria watch Sponge Bob on the Nickelodeon channel.[86]

As a form of Pan-African futurism, the Afropolitan mediascape highlights two notions: first, that Afropolitanism is a distinct experience of globalization that is enabled and made visible by new media (ICT, mobile phones, the Internet, digital media, cybercultures); and second, that as a social movement, Afropolitan lives reveal contiguous and competing experiences of globalization that reaffirm Appadurai's notion of pluralistic global flows—a collectivizing and variable "landscape" of social experience for networked and deterritorialized communities in the network society, as opposed to a seemingly homogenized public sphere (Habermas), unified *agoras*, or digital commons, as often identified in the early era of social media. Media are central to these flows. For Appadurai, *mediascapes* represent one of many multivalent flows experienced in a transnationalism enabled by information systems such as the Internet. This is similarly a repertoire or range of techniques that, taken collectively, allow for the global emergence of culture via increasingly "real-time" encounters.[87]

Theorizing cosmopolitanism, Kwame Anthony Appiah provides an expansive description that embraces the concepts' moral, cultural, and political capacities: "It begins with the simple idea that in the human community, as in national communities, we need to develop habits of coexistence: conversation in its older meaning, of living together, association."[88] Yet cosmopolitanism is also a dream or ideal, a label that those who strive place on a place, and on the residents, the individuals in those places, as if to mark each unit an unconscious member of a larger whole—nodes in an obvious and invisible network of commerce and civilization. It is an idea that strives to be "global" but is just as likely "European" and subject to its own metropolitan parochializing. According to the airplane magazines, it is Victorinox, the Louvre, the Hague, and *Cosmo* magazine; New York or Washington, DC's cosmopolitanism is uniformly assumed. Yet in these acculturated spaces, we see *a kind* of globalization, that is, transnationalism, of a kind. What is the cosmopolitanism of Manhattan, NY, a global center of corporate finance, fashion, culture industry, and media, compared with Washington Heights, a global Latin Caribbean neighborhood in the same borough? How should we characterize the "Little Africas" of the Bronx, Houston, Minneapolis, or Naperville, Illinois, just outside of Chicago? By contrast, the cosmopolitanism of these spaces is found in the churches, night clubs, markets, and homes in suburban Maryland or Columbus, Ohio:[89]

locales for family gatherings, fundraisers for homeland politicians, sites for music videos, and sets for Ghallywood films. Like the nightclub or church, these are spaces of pleasure and social imaginary, where taxi drivers, hair dressers, and doctoral students similarly circulate. Whereas cosmopolitanism seeks a kind of universal set of codes to be recognized, the Afropolitan ethos collectivizes those sentiments via routes that lead to Africa.

In these mediascapes, networks do matter. Consider the various Afropolitans discussed in this chapter: a Ghana-born musician based in New York who regularly tours Europe; the multisited author, Harvard educated, making one of her homes in South Asia; and broadcasters and media personalities shaping news from Johannesburg, London, or Miami. Consider the web developers based in Tema whose weekend projects are get-out-the-vote campaigns across the country, whose business contracts are in Benin and Liberia. Consider the cosmopolitanism of music stars culled from throughout Ghana, even the United States, sporting shorts made by Dutch clothier Vlisco in styles cribbed from local African fabrics. Reggie Rockstone, who returned from diaspora to start the hiplife rap movement in the 1990s, at one time dubbed his Accra club Django, a reference to the 2012 film about a vengeful ex-slave starring Jamie Foxx set in the pre–Civil War American South. How shall we characterize the Ghanaian truck driver whom I interviewed living in San Francisco (see chapter 3)? Ahmed formerly made a living selling, transporting, and fixing motorcycles throughout Togo, Benin, Nigeria, and Cameroon, before moving to the US. When his children were young and living in Accra, he called nearly every night for 10 years via mobile phone and using long-distance calling cards to where they lived in Nima, Accra. If we can now use the *network* metaphor to map social relations, we must ask, what relationships define the networks? How are the nodes arranged, not only geographically, but functionally and relationally? How does *flow* define these relationships? What is meant by a global network, especially one like an Afropolitan mediascape(s), to a transnational or a community of them, where hierarchies persist, but in which pop culture discourse attempts to collectivize sentiments of freedom and agency? As Ashong said to me, "Some hiplife probably is Afropolitan, some maybe not. Some Nollywood is; some, not so much."[90] These are particularistic and also universalizing spaces. Nkechi Asante, a Ghanaian Dutch concert promoter I spoke with in 2022, described the emergence of the populist Afrobeats music style as Afropolitan in this way: "I can't even recall how we used to call [the music] before. It was just Ghana music and Nigerian music. There was a gap, and there was always a Ghana versus Nigeria kind-of-thing, but Afrobeats has brought us together."[91]

Afropolitans seek to elide regional and legal distinctions in favor of a greater social imaginary of Africanness unlimited by geography. As illustrated earlier, Ghanaian tech innovators move in this Afropolitan mediascape. Their projects are similar works of representation and social transformation, moving beyond a speculative future. Their work reassembles identity through digital diaspora, against asymmetrical networks, and attempts to fashion a new global imaginary for Ghana and the continent as a whole through the practices of innovation. Against Europe's cosmopolitanism, which assumes an easy instrumentality in the network society, Afropolitan mediascapes are among the increasing diversity of contemporary information networks, but ones that from a Pan-African futurist frame seek to reassert Africa in the world.

CHAPTER 5

Pan-Africanism or Africapitalism?

The Paradox of Entrepreneurship as Liberation

STITCHING TOGETHER A FUTURE

When I received the call from Kofi Marfo,[1] I had barely woken for the day. I was attempting to kickstart my brain by listening to the news on my yam phone's FM radio receiver. It was November 2016, and Ghana's presidential elections were well underway. "Good morning, Reginold. I am coming with my driver to pick you up," said the young man, aged 27 at the time. I could hear a rush of air and street noise in the phone receiver as the car sped to meet me in the Osu neighborhood. Kofi's voice also indicated the need for haste. "We'll go over to my factory, but I have an errand first." Now fully at attention, I had forgotten about our appointment, or rather was not exactly sure if it was happening or why. A mutual friend from Berkeley insisted I contact her younger brother in Ghana, "If you're studying innovation, you must meet my brother Kofi, he's into fashion. He's really a character." Kofi is a clothing designer and event planner, among his other interests. In Ghana, I had spent the last several years meeting software developers, emerging social media influencers, and gregarious social entrepreneurs with start-ups and tech schools, focused on spreading the mantra of ICT for development with a Pan-Africanist zeal. I wasn't sure what I might learn from someone whose interests lie principally in runways, fabrics, and buckled shoes.

A compact sedan pulled up in front of my apartment, its tires aggressively spinning on the gravel a few feet in front of me as it stopped. We quickly exchanged pleasantries before I got into the car. Kofi said he wanted to show me his family's property; the building was part of a compound of several apartment buildings and a newish home in the Dome neighborhood on the city's northern edge, which his father, a retired professor, had built up during his years abroad in the US. As we drove, I asked him about the Internet in Ghana, as a way of making small talk. He said pointedly at the time, "It sucks. Africa Online is always down, Main One is a bit slow," and quickly changed the subject, pointing to the luxury apartments and office spaces transforming Ghana's skyline. "Oil money," he ventured from the front seat. Speaking through the wind, he returned to his favorite subject, passing me his smartphone to show me a runway feature he had recently organized, pointing out the logos on his shoes and button-down tunics.

At some point I lost track of our direction, but it was clear we were heading toward Nima, Accra's famously bustling *zongo* with its collection of markets and migrants from Northern Ghana and French-speaking West Africa. Kofi's driver pulled into what seemed to be a dry riverbed along a major road. We passed by a series of open-air shops selling mechanical parts. In one of those stalls, I noticed a man stitching clothes using a sturdy little black sewing machine that looked as if it were 100 years old. I had seen many of these "Singer"-style devices during my trips to Ghana, especially at the universities, and was always struck by their age and seeming indestructibility; their slender design and flywheel reminded me of train locomotives. We pulled up to a hectic corner between several small used-car parts markets, where a man named Jamal sat in front of his truck, a 1980s or 1990s Toyota. Jamal was a man of few words, for me at least. His eyes studied us, and he was mostly annoyed we were late, maybe as much as an hour by someone's estimate. The flatbed was filled with large brown boxes with red lettering on the sides. Kofi and Jamal greeted each other with a handshake and snap and began to discuss the purchases: a cache of sleek new electric sewing machines newly arrived from China. It seemed for some, tailoring was leapfrogging into the 21st century.

Kofi complained a bit after opening and inspecting the contents of the boxes: there were supposed to be seven of them, but there were only five. The seller was adamant at first that he was entitled to the price for seven. It was unclear whether or not two more would arrive. Kofi, playful but firm, insisted that he would only pay for the five. Jamal shifted the toothpick between his teeth and grudgingly consented, from what I could tell. Most of the exchange was spoken in Pidgin and Twi. Satisfied, Jamal held his phone and asked, "Tigo or Momo"—the payment was to be made using a digital wallet. I was expecting a rather tedious handover over of a few thousand cedis, but instead it would be the first time I witnessed such a large exchange via mobile phone in Ghana. Standing there, amid Nima's dusty roads and ephemeral structures, it became obvious to me that a revolution in digital finance was taking place in this small corner of an African market.

. . .

In this final thematic chapter, I examine the discourse and practice of entrepreneurship as a means of economic and political liberation among activists, developers, and everyday people in Ghana. Entrepreneurship in Ghana, in Africa, and the rest of the Global South is now being advanced as the primary route toward economic development. Entrepreneurs are cast here as champions of development, especially those operating from what is often termed the "bottom of the pyramid"—a deficit perspective on income inequality, describing the three billion people across the global who live on or below US$1 to $3 per day.[2] While some have characterized benevolent approaches to market inclusion and wealth redistribution as *social entrepreneurship*, I begin this chapter by providing a critique and analysis of *Africapitalism*, which in many ways philosophically serves as a metonym for projects that use technology and business innovations to drive social change in Africa. Entrepreneurship and the discourse around it—especially via socially conscious start-ups, microenterprises, and microfinance—has had a tremendous impact on the discourse of development as of late, shaping policy of international aid organizations, foreign NGO work, and government development agendas in Ghana, just as technology start-ups in the Global North are lauded for their ascendance as some of the richest companies in the world. The central thrust of Africapitalism is that market forces, having more impact than governments on the continent, should be prime producers of "social wealth" as they accumulate corporate profits. Social wealth is ill defined, however, and in practice often manifests as philanthropy. The increasing neoliberalization of entitlements in Ghana, typified by reactions to the #FixtheCountry movement, and the hazy discourse of economic inclusion from Africapitalism highlight the limits of a politics (or anti-politics) driven by the argument for economic autonomy alone. As I illustrate in the following sections, it can be somewhat ironic then, that the label "Pan-African" is increasingly tied to the transnational ventures of African entrepreneurs. Through these arguments, it will be clear that an investment in commerce and technology devoid of its humanist roots works against the agency of Pan-African futurists. Instead, such a political economy can lead to what some have described as "neoliberal democracy."

ENTREPRENEURIAL LEADERS

In the narrative of Africa Rising, entrepreneurs such as Kofi and Jamal are cast as not only savvy operators navigating an increasingly accessible world economy, but economic heroes and central to 21st-century nation building on the continent. In such rhetoric, entrepreneurism is seen as the central mechanism (if not ideology) required to drive wealth creation and jobs as the most important outcomes of socioeconomic development. In fact, many argue that entrepreneurship has been an integral part of the formula that explains Africa's moment of spectacular growth between 2000 and 2015, which saw an increasingly favorable business environment fueled by rising exports in the oil, mineral, and natural resource market; the rapid adoption of mobile telephony and digital media; and rapid expansion of the middle class. For neoliberal prognosticators such as Dambisa Moyo, it is entrepreneurs who have been and will be essential to sustaining the rise,[3] allowing Africans to pull themselves up by their bootstraps. As Moses Ochonu explains in a pioneering edited volume on the subject, "The figure of the entrepreneur has emerged as an organizing idiom for articulating the economic hopes and aspirations of various African societies. . . . [They] are collectively regarded as the vanguard of a new African economic and developmental order."[4] The World Bank has been persistent on this point: in 2012, with the release of its Entrepreneurship Database, the organization signaled its paradigmatic interest in the topic, proclaiming, "Entrepreneurial activity is a pillar of economic growth." Tracking its growth, particularly in new companies started by women and young people, the Bretton Woods institution argued, is essential for understanding the relations of job growth and increasing wages in the developing world.[5]

Following our purchases in Nima, Kofi, his driver, and I returned to his family estate in Dome, where immediately upon arriving, a group of four additional young men came to greet us. With few words, they unloaded the truck and began setting up the new electronic sewing machines in a room in a building adjacent to the Marfo house. This was the "factory" that Kofi had spoken of, and within an hour's time all five men (including the driver) were spooling thread and testing out the new machines with lengths of cloth. They were earnest and focused on learning the new devices; more employees had been added to the Ghanaian economy that day, and as Kofi described, everyone would also receive their pay via mobile wallets.

For many, financial exchanges via mobile money, like these, are thought to be the core technology driving African economic transformation, and they are one of Africa's most singular technical innovations in the last 20 years. As the mobile phone "explosion" overtook the more industrialized and urban centers of the continent in the early 2000s, pay-as-you-go credits had become the typical means of financing mobile telephony and wireless Internet access among African tech users. For the majority of these users, consumer credit and billable subscriptions were often unavailable. Relatedly, the label *unbanked* has been applied to as much as 80 percent of these users across various African markets[6]—a description of people who typically lack easy access to financial institutions and ATMs, credit or debit cards, or loans and financing. M-PESA, a wireless person-to-person payments method using simple text prompts on smart and feature phones in Kenya,[7] became the market leader that solved this problem around 2007. Started by Kenyan and British developers, M-PESA's inventors at Safaricom, a local telecom, essentially digitized the process of informal credit exchange that had been taking place in the early years of Africa's mobile adoption. With paper credits purchased from retailers on the street, mobile phone users entered access numbers to "top-up" their minutes, through a series of text-based alphanumeric transactions.[8] Very often users would pool paper credit purchases and split the minutes between their accounts. Often the telecom providers' paper "top-up" tickets would serve as a de facto form of currency. By 2007 Safaricom had formalized the process, using smartphone apps to access digital wallets, and importantly had extended access to feature phones through a text-messaging menu called USSD. Simple feature phones remain the dominant mobile devices on the continent even into the 2020s, and this vernacular design was a tremendous enabler for M-PESA. The procedure of using SMS or "short codes" to load credit became a new market model. These forms of exchange allowed users to make transactions and payments without the need for physical paper, money, and sometimes even banking accounts. Once more, digital wallets largely skipped the use of formal banks, as the telecom companies were granted the ability to develop financial divisions in markets such as Kenya, Tanzania, Uganda, and Ghana. Mobile money had done what 50 years of economic development and banking growth had not: extend electronic forms of credit and exchange, including Internet-based transactions, to everyday users without the use of tools such as credit cards. This provided for enormous social transformation in the parts of the continent where access to banks is plagued by their geographic centralization in cities, the existence of few branch offices, itinerant electricity and inadequate civil infrastructure, and literacy and cultural challenges at banking institutions, not to mention extremely conservative bank policies and high lending rates for consumer credit. Mobile wallets became accessible as mobile phones became more plentiful, and the practices vernacularized according to the legal requirements and local uses in each country. These forms of digital finance have allowed products such as microfinancing, insurance, and savings to develop as well.[9] Such payment methods and platforms, in which the user is required neither to use a smartphone nor to have a formal bank account, have transformed commerce on the continent in the last 10 years. As of 2024, mobile wallets had only just begun to bridge transnational systems, with homeland remittances constituting as much as 4 percent of their transaction volume (US$29 billion).[10]

In 2016, when I witnessed the transaction over the sewing machines, mobile money was still just getting off the ground. At the time, figures from McKinsey & Co. numbered mobile payment platforms in Africa south of the Sahara at 143 firms—more than three times that of Asia, then the second-largest world region for digital wallets. Estimates were that the value of mobile financial transactions would soon exceed US$2 billion.[11] Three years later, African mobile electronic payment revenues were tallied at more than US$19.3 billion.[12] The NCA reported then that the value of transactions totaled nearly 90 percent of all the funds held by customers in the Ghanaian banking system.[13] The Nigeria-based Jumia, an online market space similar to Amazon, grew to be Africa's first "unicorn"—a term marking more than US$1 billion in company valuation. Jumia achieved this on the strength of its use of e-commerce and mobile wallet transactions, and was soon joined by other fintech companies such as Nigeria's Interswitch and Egypt's Fawry. Flutterwave, founded in Lagos and now headquartered in San Francisco, claimed to have processed more than US$2 billion in payments in 2022 alone.[14] Backed by venture capital from the start-up incubator Y-Combinator, Visa Ventures, Mastercard, and Microsoft, Flutterwave has become among the richest and most recognizable platforms for digital wallets and mobile money payments. These successes, touted by developers in Ghana, on popular African-tech news sites such as TechCabal, and on podcasts such as *African Tech Round-Up*, have solidified for many the potential in the popular narratives of an "African Dream"—a vision of future personal and national economic stability and growth for the continent driven by ICT innovation.[15]

In spite of the growth in commodities and oil exports in the last decade and the expansion of the telecommunications industry with continent-dominating firms such as MTN and Sudanese billionaire Mo Ibrahim's Celtel, it is in many ways the disruptive phenomenon of mobile money that has attracted Western investors to African's consumer markets as of late. While the debates about the efficacy of Africa Rising have persisted since 2010,[16] the growth in digital wallets has served for many as evidence that African economies are indeed viable sites of free-market globalization. But what are we to make of this highly lauded expansion of capitalist enterprise on the African continent and its accompanying rhetoric of economic liberation through private enterprise? In its genesis, Pan-Africanism was conceived as an anti-imperialist, political movement, organized against racial capitalism and often socialist or Marxist in its later manifestations. How can Africapitalism also be Pan-African?

WHAT'S AN AFRICAPITALIST?

The ability to use capitalism as a tool for African autonomy is an aim often stated by many of the activist-developers I encountered during the course of my research in Ghana. When I was chatting with Ato Ulzen-Appiah in 2023, the director of GhanaThink said to me, "I love Nkrumah, but I am much more of a capitalist than a socialist."[17] For many social entrepreneurs in Ghana, to be Pan-African is not (simply) about political unity or independence, but also, as my research contact Victor K. Ofoegbu from Impact Hub Accra asserted, Africa's ultimate "economic liberation." It is for many the "scaling" of enterprises across African borders using tech. As Ofoegbu and many others relayed to me, "For us, it's the last stage of the independence movement."[18]

The rhetoric that entrepreneurism is often described as the key means to "financial liberation," found among many of my interlocutors, is also evident on Ghanaian social media, in podcasts on technology in Africa, and at the growing number of business-oriented African development conferences at universities throughout the United States. Entrepreneurs are cast as the continent's new "disruptive" leaders at meetings such as Harvard's 25-year African Business Conference and newer affairs such as the African Diaspora Networks' annual meeting in Silicon Valley,[19] along with similar conferences that have emerged in Europe, China, and the Middle East in the past decade. At these events, and in the African mediascape more broadly, the contemporary desire to theorize and utilize capitalism as a tool for development and "economic empowerment" is often cited as *the* missing component of the push for African independence during the 1960s.[20] While development agendas emerging from these business and technology summits rarely cite the similar programs of socialist "modernization" deployed by new African governments during the period of independence, the term *Pan-African* or *pan-African* has appeared more frequently as a catchall to describe the transcontinental outlook for these enterprises. Africapitalism, while not universally used as the moniker for African and Ghanaian entrepreneurial-focused development, serves as a useful metonym from which to interrogate the ethos of what Jemima Pierre and others have described as "neoliberal Pan-Africanism."[21] These efforts have embraced not only free markets, but a Pan-African futurist sensibility, as technology and start-up enterprises are central to their tactics for social change.

Since the 2010s, the notion of "dead aid," coined by the economist and writer Dambisa Moyo regarding the failures of donor-funded development, has continued to resonate among aspirant youth in Africa and was everywhere in Ghana during my years of research there. In Moyo's formulation, the past 50 years of Western charitable donations, crisis funding, and state-sponsored development grants (e.g., USAID, UKAID) have not helped Africans in making their societies more resemble the West; rather, foreign direct investments and charity have "slowed its [Africa's] growth." NGOs and the work of groups like CARE International, as she states, "make the poor poorer."[22] The solution for African governments is to develop local entrepreneurs who can seek investments and partnerships with multinational corporations to generate growth and jobs and raise standards of living. In many ways Moyo's thesis, supported in the popular press and via venues such as the World Economic Forum, reflects the critique of dependency theory, repackaged for the 21st century. The central theme in Samir Amin's formulation of "unequal development," and in the work of critical *dependentistas* such as Fernando Cardoso and Enzo Faletto,[23] is that the impoverishment of the Global South has been directly proportional to the West's global domination and wealth accumulation. Leftist critics of economic development programs for Africa have argued that whether through mercantilism, imperialism, colonialism, or programs such the World Bank's SAPs of the 1980s (critiqued as "neocolonial"), Europe and the United States have continually made the Global South a source of cheap labor. Underdeveloped economies have been made overreliant on the export of raw materials and cash crops, further indebting poorer countries to their former colonizers while opening them up to new ones, like the US and China. Whether through the historical Transatlantic Slave Trade, economic sabotage, trade sanctions, or the unfavorable export terms of contemporary globalization, Africa has never enjoyed the kinds of external trade surpluses with the West that the West enjoys in its relations with the continent. Furthermore, even where African nations have gained politically and economically from the export of resources such as oil, gold, diamonds, cocoa, and coltan, the supply and prices for these profitable local exports have been subverted by loan arrangements with the IMF, World Bank, the Export-Import Bank of China, and organizations such as the Paris Club. Operating from deficits and now competing in a global market in which African exports are disadvantaged and susceptible to foreign competition, highly indebted nations fail to accrue enough reserves to advance their industrializing and "modernizing" goals that would enhance political and economic autonomy.[24] Both groups also place blame on what Pierre has observed as opaque and concrete instances of "bad governance."[25]

Marxists such as Amin have argued for nationalist policies that "delink" commodity flows away from Africa to former colonial powers, steering these toward more mutually beneficial South-South trade and development schemes with Asia, the Middle East, and South America, for example; Moyo and other neoliberal theorists, however, have been arguing for an entrepreneurial-led revolution in Africa,[26] utilizing ICT to drive service-sector growth that creates local firms to partner with multinational corporations. This confluence of export nationalism and regional opportunities for trade and partnerships via multilateralism through ECOWAS or the AfCFTA has driven Africapitalist advocates to develop a continent-wide perspective on growth,[27] using the rhetoric of national liberation and Pan-Africanist sentiments. Paired with the libertarian "Californian ideology" of the tech industry,[28] the label "Pan-African" has increasingly become less an anti-colonial political identification (in the ways described in the previous chapters) and more a way to describe market opportunities for entrepreneurs and emerging businesses across the continent.

In public lectures and TED Talks, on podcasts, and in the foreign and homegrown NGO community, such neoliberal, anti-statist positions have reframed the continent as an untapped marketplace, "the Bright Continent" as Dayo Olopade has described it.[29] Standing in the way, say many, is bad governance—postcolonial bureaucratic malaise and corruption—an at times fair assessment,[30] that also largely undergirded the principles of economic development central to the aims of SAPs in the 1980s and in the time since. This new iteration of Pan-Africanism is less strident in its critique of Europe's continuing legacy of colonial relations or even of Africa's current unequal relationships with China, the Middle East, and others.

As a movement, Africapitalism remixes 20th-century notions of Pan-Africanism born in the liberationist discourse of figures such as Edward Blyden, Léopold Senghor, Kwame Nkrumah, Patrice Lumumba, and Queen Mother Moore, among many others. German researcher Tim Weiss described this spirit in *Digital Kenya* in this way: "In particular, Pan-African conferences, workshops, Internet platforms, and organizations create a new dimension of entrepreneuring, one that works to overcome and remove constraints at the transnational level."[31] As developmental economist Rita Kiki Edozie points out, emergent African market research organs such as the Pan African Market Growth Index reveal "Western appropriation of African 'Pan African' self-determination and oppositional narratives and tropes."[32] Yet as procapitalist Nigerian podcaster Dotun Olowoporoku, host of the tech show *Building the Future*, described in an interview with me in 2018: "Pan-Africanism is seeing beyond the drifting geographic barriers that was drawn up by some old White people over a 150 years ago [sic]. We have opportunities beyond our borders that are out there for us."[33]

The term Africapitalism emerged in the same moment that Africa Rising rhetoric began to take hold in the new millennium. Heralded by the successful Nigerian banking magnate Tony Elumelu, the concept Africapitalism was described in a series of talks and white papers, touting a new "management idea for business in Africa" in which "Pan-African" exchanges would be at the forefront. As Elumelu stated in the inaugural issue of the newsletter *The Africapitalist* in 2010: "True growth and economic development in Africa will only be achieved with private sector commitment to economic transformation—one that seeds new businesses, puts economic well-being within the reach of millions, and helps solve our most pressing social problems. We call this Africapitalism™. Africapitalism™ is an economic philosophy that embodies the private sector's commitment to the economic transformation of Africa through investments that generate both economic prosperity and social wealth."[34]

Since 2010, the Tony Elumelu Foundation (TEF) has branded this philosophy, which Elumelu has promoted alongside the vertical growth of the United Bank of Africa and the Heir Holdings private-equity group, which he helms. Elumelu's rise began in the late 1980s, after graduating with economics degrees from the University of Lagos. In 1997 he led a group buyout of Crystal Bank, renaming it Standard Trust Bank. In 2005 Standard acquired United Bank of Africa in what is often described in its literature as "one of the largest mergers in Sub-Saharan Africa at the time."[35] (Elumelu has also described the venture as "democratizing African banking.") In the following years, the UBA brand became ubiquitous in Nigeria and grew interests in 20 other countries on the continent. In 2011 Elumelu also started the Africapitalist Institute, now the TEF, and began promoting a benevolent approach to socially responsible business centered on "social wealth." In a manner not unlike the US State Department's Mandela-Washington Fellows or the Meltwater Institute in Ghana, TEF launched an entrepreneurship fellowship program to train cohorts of young business professionals from around the continent on its digital networking site, TEF Connect. Functioning as a tech innovation hub and business incubator, the entrepreneurship program claims to have invested over $100 million in 18,000 program participants to create new businesses and jobs. (*Forbes* would briefly name Elumelu one of the richest men in Africa in 2015.) In the same inaugural *The Africapitalist* newsletter in 2011, Elumelu wrote that UBA had grown into "a pan-African financial services company operating in 20 countries across Africa as well as the Middle East, Europe and the United States."[36] Upon UBA's stock market valuation of ₦1 trillion (US$666 million) in early 2024, the company's CEO, Oliver Alawuba, remarked, "Market participants have begun to appreciate the latent capacity in UBA's business model as the bank unlocks enormous potentials in its pan African [*sic*] and international operations."[37]

Elumelu and other business thought leaders in the past decade have been particularly focused on extolling the virtues of African private enterprises as being key to the growth of the continent's economy since the 2000s.[38] These narratives emerged just as the Jubilee 2000 movement advocated wiping out decades of African debt owed to organizations such as the IMF and the Paris Club,[39] and as African nations arguably began to stabilize from the liberalization of their national economies, including the privatization of some state-run enterprises and utilities such as telecom and electricity.[40] The end of the "Structural Adjustment Era" seemed to provide African nations with a new start and also coincided with the fall of the Soviet Union and the ascendance of democratically elected leaders on the African continent, notably in Ghana (1992), South Africa (1994), and Nigeria (1999), just as the era of digital globalization (the network society) emerged as the dominant global system.

AFRICAPITALISM'S MORAL IMPERATIVE

Africapitalism may have emerged as a slogan for rethinking the role of the private sector in "financial returns to shareholders as well as economic and social benefit to stakeholders."[41] But its intellectual heft was cemented with a series of TEF-funded workshops in 2014, coconvened by management professors Kenneth Amaeshi (University of Edinburgh and Lagos Business School) and Uwafiokun Idemudia (York University in Toronto). Two subsequent academic edited volumes followed: *Africapitalism: Rethinking the Role of Business in Africa* (2018) and the more theoretically balanced *Africapitalism: Sustainable Business and Development in Africa* (2019). While chairman Tony Elumelu has clarified the notion since 2011, it is in these collections (more than 20 chapters) that the "management philosophy" is considered more robustly. Across these theoretical texts, Amaeshi and colleagues' examination of the potential for an Africapitalist developmental framework is distilled into four central tenets.

Describing the business slogan as a "call to action" and an "imaginative moral-linguistic project," the authors and edited volumes assert that an "indigenous" capitalism for Africa should possess a "Sense of Progress and Prosperity," a "Sense of Parity," a "Sense of Peace and Harmony," and a "Sense of Place and Belonging." Introducing these principles, state Amaeshi and colleagues, "Africapitalism is predicated on the creation of social wealth in addition to the pursuit of financial profitability."[42] The authors state that in order for progress to occur, conditions must emerge that "make life more fulfilling" on the continent, a statement that calls to mind the central tenet of humanitarian-focused development ideas advocated by the likes of Nobel Prize–winning economist Amartya Sen. Sen famously defined *development* as a sense of *freedom* that "enables people to life they have reason to value."[43] Rather than anchor Africapitalism within this framework of developmental economics, Amaeshi, Idemudia, and many of their interlocutors draw from the literature of corporate social responsibility (CSR),[44] in which ethical costs of wealth accumulation are weighed against the production of surplus wealth. Usually the values of profitability win out.[45] In Amaeshi and colleague' description of their more egalitarian notion of "parity," the authors, echoing Elumelu's sentiments, state that Africapitalism must "promote a form of entrepreneurship that strives to create financial and social wealth for all stakeholders and not just for shareholders."[46] The principles of peace and harmony are couched as "sustainable development" principles, noting that equity should be central to Africapitalist outcomes given "the tendency of liberal market capitalism to lead to some form of socio-environmental imbalance."⁴⁷ A "sense of place" resists the flattening of sociocultural space (Africa) as an economic opportunity within neoliberal market discourse, engendering instead among Africapitalists a sense of economic and "corporate patriotism." This suggests that "the firm be seen as a community rooted in a sense of place."⁴⁸

In these descriptions, notions of Africapitalism as a "linguistic project" or "philosophy of management" seem to highlight the great rhetorical efforts Elumelu, Amaeshi, Idemudia, and their interlocutors work through in order to avoid using terms like *ideology* in their writing. As a set of principles and a moral ethos, one must draw attention to the soft ideological claims inherent in this "philosophy" as a vernacular theory, in much the same way that other "worlds of capitalism" (as they describe China's policies, for example) have emerged. What Africapitalists may be rightly intuiting is that such a cogent, mobilizing agenda requires ideological thinking that signals a kind of politics—in this case, the continent's reappropriation of neoliberalism. Clearly, central to Africapitalism is a call for developing an ethical framework that supersedes notions of individuality, ethnic allegiance, and nationalism, and gestures toward a perhaps internationalist predisposition with regard to the continent's economic development. China's "collectivist" ideology (communism) and its development strategy of "social market capitalism" is often cited by Africapitalists in the two Amaeshi-led texts, as an analogous approach to development, similarly focused on producing broad social wealth based on communalist values. But what remains ill defined and inadequately discussed throughout the pro-Africapitalist literature is the notion of *social wealth,* as noted in critiques of the idea, including by Stephen Ouma, whose work appears in one of the volumes. As he states, "Attaining the power to shape one's own destiny and developing a set of discursive, place-based concepts . . . are certainly key to a more prosperous African future. It can be questioned whether this should be done through practices that have historically built wealth in certain regions of the world only on the back of cheap nature, food, labour and energy elsewhere."⁴⁹

The linguistic nods to Pan-Africanism here bring the Africapitalist idea into more focus as a movement and as a broader ethos that link advocates for Elumelu's entrepreneurial-led initiatives in Nigeria to regions across the continent in general, including Ghana's activist-developers and social entrepreneurs: *Pan-Africanism is the central ideology that Africapitalism operates under.* In advocating for transcontinental solidarity, an end to systematic anti-Black racism, collective advancement, and a futuristic march toward "one destiny," Africapitalism utilizes Pan-Africanist ideals, paradoxically advocating for market-driven solutions to the enduring issues of postcolonial political economy and autonomy.

Amaeshi and colleagues state emphatically that "capitalism is primarily a moral project," and from the Africapitalist's perspective, one "rooted in the values of Ubuntu, both as a process and as an outcome."[50] Embracing an African-centered, cosmological sensibility, the concept of *ubuntu*, the Southern African social principle of collective identity often interpreted as "I am a person through other persons" or "You are a person because we are persons,"[51] is often invoked within Africapitalist discourse, in fact used more than 100 times in the 2019 Amaeshi and colleagues edited volume. In her 2017 book *"Pan" Africa Rising*, development economist Rita Kiki Edozie centers Africapitalism alongside notions of "Ubuntu Economics" in her exploration of Afrocentric, market-based development strategies, where she profiles the work of South African thought leaders such as Mfuniselwa Bhengu, the business developer Reuel Khoza, and Tony Elumelu. Yet such "philosophies" are also potentially antithetical to principles of *ubuntu*. As technologist and philosopher José Cossa states, *ubuntu* implies that "our deepest moral obligation is to become more fully human, and to achieve this requires one to enter more deeply into community with others."[52] While Edozie and others wax enthusiastically about the potential for socially responsible economic approaches to provide a "third way" for the continent,[53] we must pause to consider how, as Ouma and others have noted, foundational approaches to capitalism in Africa have been imperial projects, in which competitive scarcity and monopolization are the hallmarks of its resource distribution schemes— often at the expense of indigenous Africans. What is more potentially transformative and also vexing within the Africapitalist framework is the way that "social wealth" is invoked as the ultimate goal. In a TEF white paper dated June 2022, Elumelu states that African business leaders "must break free from the historical tendences of exploitation and extraction and instead focus on generating profit through wealth creation." He later expands on this, simply to say that wealth creation by the private sector is "the most urgent priority."[54]

Stated plainly, the notion of social wealth is highly undertheorized in most Africapitalist discourse, and so more poignant questions must be asked: How is wealth, or social wealth, to be defined? What qualities of environmental or social sustainability and equity are built into the Africapitalist sensibility? Is there a practice of revenue sharing or cooperative ownership or other collective economics? How are we to distinguish this idea from the notion of *social welfare*, traditionally embedded in the state policies toward citizen entitlements, especially in formally socialist republics? Is there an aversion to such an idea and thus a deliberate attempt to evade such comparisons?

Through the persistence of notions of social wealth, one sees a connection to the NGO drive for *social entrepreneurship* as a developmentalist project.[55] Social enterprises, including many of the projects discussed by activist-developers in this book, insist that both personal and social development can be achieved through profit-making ventures, in which some financial gains are, for instance, donated to charity, or that the paid-for services themselves are socially redeeming. Attached to practices such as social entrepreneurship or entrepreneurship in general, personal profit seems to have a premium above collective notions of resource (re)distribution. In Ghana, beliefs about material accumulation and the good life are also tied to popular discourses such as that of the "prosperity gospel" of megachurches, in which social wealth is coproduced through religiosity and spiritual favor.[56] Here again, wealth is evidenced as a material surplus. Even with such descriptions of capitalist wealth generation as having a broad social benefit, one fails to see how accumulation manifested at the individual rather than social/collective-level squares with an ubuntuist perspective on social welfare, in which resources are equitably distributed both systematically *and* in the lives of individuals.

Ironically, moreover, the drive to "modernize" Africa in the 1960s with state-owned enterprises (e.g., Nkrumahism in Ghana or *ujamaa* in Tanzania) were similar experiments in growth through industrialization, with much the same aims of economic expansion, though invested at times in all the trappings of and indeed collusion with capitalist enterprises. In Ghana, projects such as the Akosombo Dam and the Port of Tema were techno-futurist developments, whose aims were at once decolonial and intended to assert international autonomy through managing national resources. The dam's construction would provide a system for electrification in the southern and central regions; the state-run ocean docking systems in Tema, now the third largest in West Africa, centralized shipping and trade in the region.[57] Similarly, the Ghana Atomic Energy Commission and the 1964 Seven Year Plan for National Reconstruction and Development are remarkable in this sense, as both these projects promised developmentalist modernization, delivered through an agenda of science and engineering in ways similar to the Pan-African futurist aesthetic. As Osseo-Asare states in her history of the nuclear energy reactor project in Ghana, Nkrumah envisioned "cadres of elite scientists" who would "amplify the industrial ambitions of the new nation, perhaps even propelling them to space someday."[58]

These developments were the product of state planning inspired by Soviet Russia rather than free-market enterprises, and thus were rooted in a socialist ethos of collective wealth, analogous perhaps to today's notions of the peri-capitalist, developmentalist state.[59] Still, as economic historian Kwadwo Osei-Opare helps us to see, the notion of capital-generating enterprises, including partnering with private industry, was central and not peripheral to the Nkrumahist plan for modernization and industrial development. Oakland's Kaiser Aluminum Co. (problematically) was at the center of the Akosombo Dam project and plans for an accompanying aluminum smelting factory. In 1963 Ghana lifted a requirement that 60 percent of profits of foreign firms be reinvested in the country. Osei-Opare quotes Nkrumah at the opening of a Unilever-run soap factory in Tema as stating, "Some people think that Capital Investment is in contradiction with our socialist aims and ideas. This is not true."[60]

Pan-Africanist leader Marcus Garvey's "Black capitalism" strategy is an important precedent to Africapitalism and important to note here. Pan-Africanism as a cogent philosophy started in the Caribbean in the early 20th century and grew with adherents in the United States, and then Europe, where diasporic Africans were organizing against colonial rule in the homeland. Garvey advanced a political ethos of nation building and an economic model that emphasized commercial enterprises, in ways similar to earlier advocates of self-sufficiency such as Booker T. Washington. Yet because of his Pan-Africanist political agency and rhetoric, Garvey's promarket sentiments were largely elided by the US government, which saw him as a threat to the transnational racial order of White hegemony.[61] In the early 20th century, amid differing political organizing tactics and economic visions of Black autonomy (Liberia, Ethiopia, African Black Brotherhood, Tuskegee Institute, Négritude, etc.), Pan-Africanism provided a binding philosophical association against European empire, and as Burden-Stelly describes, organized struggle against "capitalist racism."[62] In part, this explains the attraction of so many African nationalist and diasporic independence movements to Marxism and socialism, as a moral, social, *and* economic critique. African communalism and European notions of the social contract provided the basis for an emerging anti-imperialist ethos that was core to the early theorizing of Pan-Africanism. Yet in Africapitalism, paradoxically, Pan-Africanism is sheared of its moral imperatives for economic and political justice and the African historical-cultural context, in which anti-capitalism was seen as central to independence because of the role that capitalists have played in destabilizing the continent's autonomy. Pan-Africanism was born a political strategy that emphasized not only unity but collaboration against colonialism and imperialism. If socialism, Marxism, and at the very least a sense of originary African communalism such as Négritude, *ubuntu*, or *ujamaa* were foundational to most articulations of Pan-Africanism, then an African-centered capitalism would necessarily seem incompatible with the principles of Africa's most pervading political-economic ideology.

THE PROBLEM WITH PAN-AFRICANISM AS A BUSINESS PROPOSITION

If Africapitalism as a political liberation strategy seems inchoate in our moment, a few additional episodes should highlight the potential trouble with utilizing Pan-Africanism as one among many business strategies, especially applied cursorily as a geographic description for corporate enterprise. In early September 2021 I received an email informing me of a new university in the small African archipelago nation of Mauritius. Mauritius has always been enigmatic in African affairs—its autonomy secured in part by its role as an offshore tax haven for foreigners doing business on the continent, in nearby India, and with the Middle East.[63] Mauritius's economic development success has always been a source of envy for the rest of continental Africa, going from per capita earnings in 1968 of US$200 to US$12,500 in 2020, and no longer being cast as a "developing" nation.[64] Like its close geopolitical ally and neighbor, the Seychelle Islands, Mauritius has a murky history with regard to notions of self-sovereignty. At the height of colonialism in 1896, British imperialists exiled the Asante Kingdom's leader and royal family to their protectorate, Seychelles, for more than 20 years, consolidating their rule over the central region of the Gold Coast (Ghana). During the Cold War, the United States and UK deported hundreds of ethnic Chagos from Mauritius and the Seychelle Islands in order to establish a strategic naval base on the island of Diego Garcia. Yet in the last 30 years Mauritius has gained notoriety as having the fastest broadband speeds in Africa, as the island has been a key point for landing stations of several Internet sea cables to the continent, Europe, and the Middle East and Indian region. Internet penetration is at 87 percent, with 144.24 mobile-cellular telephone subscriptions per 100 inhabitants: For many years it was the most connected country in the AU, with the highest Internet speeds. This has given this otherwise tourism-driven country a Switzerland-like appeal for African start-ups, crypto-entrepreneurs,[65] and institutions such as the newly formed African Leadership University.

ALU is the outgrowth of Ghanaian businessman Fred Swaniker's African Leadership Academy, a secondary school that started in South Africa in 2004, focused on entrepreneurial leadership. It has expanded, with tertiary college campuses in Mauritius and Rwanda in 2016. The Rwanda site has become especially significant, given the nation's growing aspirational image as an African nation that is successfully marrying investments in tech enterprise and infrastructure, neoliberal expansion, and success on the world stage, often referred to as the "Rwandan model."[66] For many, the images of Rwanda's clean streets, growing cityscape, high-tech factories for mobile phones and other appliances, and a public narrative of entrepreneurial development, has made it the gold standard for development and the envy of software developers in tech hubs in other middle-income African markets, such as Nigeria, Kenya, and Ghana. For others this "entrepreneurial orderliness" has belied an enduring autocratic political and social order that emerged in the wake of the 1994 genocide and Paul Kagame's political leadership since 2000.[67]

The ALU promotional email by Jess Auerbach from 2017, then an academic director for the college's Mauritius campus, leads with the provocative headline "What a New University in Africa Is Doing to Decolonize Social Sciences." Citing the torrent of student and race-based protests in South Africa and the US since 2016, Auerbach (herself a White South African) proclaims that ALU's work exists amid a continuum of a "much deeper history of national reimagination across Africa and the world. With this history in mind our faculty is working towards what we consider a decolonial social science curriculum."[68]

The rest of the essay is progressive if not benign, implying a Pan-African sensibility to ALU's work, in particular by calling for an African-centric approach to multitextual learning (music, architecture, and oral literature are all fair game as texts). The writer acknowledges the need to incorporate "Africa's long intellectual history" (read traditional arts and philosophy) and also training in regional languages. She argues these points in a matter-of-fact way, so as to avoid the obviously radical positioning this might otherwise engender for what has for decades been a colonial, Eurocentric reputation of education on the continent.[69] Reading critically, one wonders how such a venture, led by impact-oriented social scientists, might proceed in comparison to what is obviously a negative characterization of traditional humanist education relying heavily on theory. Will ALU adopt interdisciplinary approaches that hold at bay damaging "capacity-building" tropes?[70] Recalling the distinctions between state-sponsored African area studies programs in the US, Black/diaspora studies, and the continent's academic scholarship in African studies, should we expect interpretative work to be critical, controversial, or value added?[71] Auerbach variously incorporates classic criticisms of academia and social science for being both didactic, abstract, and out of touch with the public, while claiming to exhort students to "use the tools of analysis that they acquire in their training with real world implication." This can be accomplished with "open source" texts and research that is open to "collaboration" with corporate and institutional partners (quoting from the article here).

In this article, one is at once confronted by a paradox: How does a university steeped in neoliberal phraseology, focused on entrepreneurial development and endeavoring to work outside of politics, advance the very political idea of African economic and political sovereignty? Given such a context, how can such an organ as ALU claim that "decoloniality is central to our work"? It is difficult to imagine such aspirations to democratic humanism amid political environments in which dissidence is regularly stamped out; in the case of Rwanda, ALU's key host partner, the human rights watch group Freedom House has perennially rated the country "Not Free."[72] While Mauritius has been lauded for democratic handovers of power, the government has increasingly been criticized for limiting the activities of journalists and for harassment of media professionals.[73]

BANKRUPTING AFRICA'S INTERNET

Perhaps it should come as no surprise then that just as this neoliberal Pan-African university comes to be in Mauritius, AFRINIC, an international body governing the continent's Internet, and based on the island nation, should come under controversy. Like the global governing body ICANN,[74] the regional agency AFRINIC is charged with administering crucial IP addresses and domain name extensions for websites in the African region (e.g., .com, .org, .gov, .gh). The IP address system relies on numerical sets of codes or "protocols" that allow websites and other Internet tools to have precise identifiers from which to host pages, enable network exchanges, or interact with other media of the Internet. This is the address book of the front-facing web. In the expansion of its new

IPv6 system, which emerged in the late 2010s, AFRINIC assumed its regional authority to allocate Internet IP addresses, reserving and preserving domains for services centered in Africa for Africans, in ways thought to be equitable and benevolent. The problem arose when a lawyer from China acquired more than 6.2 million African IP addresses obtained for non-African services targeted at Asians and other users worldwide (including pornography and gambling sites) from 2013 to 2016. These domains accounted for nearly 5 percent of the continent's total number of allotted IPv6 addresses. In its coverage of the crisis, the Associated Press wrote, "AFRINIC is trying to reclaim Internet real estate critical for a continent that lags the rest in leveraging internet resources to raise living standards and boost health and education."[75]

Though acquiring IP addresses typically has an associated cost, AFRINIC and ICANN are charged with ensuring that no one entity, political or corporate, monopolizes IP addresses associated with specific domain names, as well as the numerical values for server addresses. For instance, they prevent a company like Amazon from buying all the ".us" or ".org" websites within any region. In the criminal case at the heart of the controversy, an officer at AFRINIC was accused of illegally selling 5 percent of the entire African continent's IPv6 addresses to Lu Heng, a Hong Kong–based Internet magnate, though his web businesses were headquartered in Seychelles. The AP article highlighted the transnational dimensions of this power grab, noting that an Israel-based entrepreneur had also acquired millions of African IP addresses in a similar scheme. Ultimately, these scenarios call into question the ability of free markets to advance African sovereignty, even in the digital age. Heng eventually sued and won, in the process bankrupting AFRINIC, which in 2024 sat in financial receivership via an externally appointed control board.[76] In his lawsuit against the nonprofit, Heng stated, "AFRINIC is supposed to serve the Internet, it's not supposed to serve Africa."[77]

. . .

As centers of thought leadership on capitalism in the Global South, ALU and the Africapitalist/Tony Elumelu Foundation's model of private-sector-led development have much in common with policy initiatives within the global diplomatic community and NGO world generally, where "youth leadership" is touted as an end around to government corruption. We might include Ghana's Ashesi University and Meltwater among their sister institutions.[78] These ideas were central to Barack Obama's Mandela-Washington Young African Leadership Institute, a program aimed at creating exchanges between emerging African entre-preneurs and businesses and communities in the US. Online asynchro-nous learning courses are central to these programs, and in one such YALI entrepreneurial class, the program states: "With more than 25% of the world's workforce projected to live in sub-Saharan Africa by 2050, it's imperative that young African leaders continue to create economic opportunities for themselves and for generations to come."[79]

At the center of these projects is a steadfast requirement that citizen-driven entrepreneurship should be the bedrock of any African develop-ment agenda, with advocates often citing pronouncements from the AU and others that micro-, small, and medium enterprises (MSMEs) employ the largest number of people on the continent and are the greatest source of job creation[80]—a notion that has been tempered by other research.[81] It is clear that Africapitalists seek to deploy capitalism for greater individ-ual and social freedom, and ultimately political expression, by acknowl-edging an ethos at the center of economic and social exchange that does not equate with pure market extraction. Africapitalism is instead "capi-talism with a consciousness,"[82] efforts that may be equated with largely failed or dismantled postcolonial economies and their "project of mod-ernization with an African soul,"[83] as Mkandawire has described post-independence approaches to developmentalism.

But if the ideas of Africapitalism are set to be the core of a new Af-rican economic independence movement that describes itself as Pan-African in the tradition of Nkrumah and the five Pan-African Congresses that gave ideological heft to the catchphrase in the 20th century, the con-cept of Pan-Africanism in the new millennium has come to mean some-thing very different.

In my interviews with activist-developers and entrepreneurs and in the media surrounding entrepreneurism (especially podcasts),[84] business executives and founders routinely describe a "Pan-African" strategy, which in their thinking is firmly rooted in capitalist expansion practice: the creation of more markets; the creation of markets on the continent where supposedly none existed; the financialization of these markets along Western capitalist modes of economic extraction and abstrac-tion; and the pursuit of neoliberal social, political, and economic policies that seek to monetize the most rudimentary services and obligations be-tween citizens and states. Africapitalism thus finds affinity in the Heri-tage Foundation–supported work of George Ayittey,[85] an ardent critic of African socialism, who has historically provided a very flattened analy-sis of economic structures within Africa. For George Ayittey the simple existence of broad, open public "traders markets," such as the famous Makola market in Ghana, demonstrates the capitalist leanings of African economies.[86] While Ayittey was also critical of the protectionism and monopolizing practices of the Western multinational firms and financial companies, he still advocated for the broad expansion of capitalism as a tool for social freedom on the continent.[87]

Meanwhile, wealth in Africa remains sequestered in the elite realms of power, in the space of extraction, mining (aluminum, gold, oil, coltan),[88] and the exploitation of resources, and unregulated economic outflow transfers that can further autocratic tendencies.[89] Many multinational corporations emerging out of the African continent have White South Africans or Europeans on their board of directors or as their CEOs; Africa's corporate systems remain patently colonized in some places, especially in the telecommunications sectors and shipping.[90] Even where digital innovation looks to disrupt racial capitalism on the continent by utilizing digital media as a strategic space of economic disruption, reports have emerged in the last few years of an alarming number of instances in which the majority and lead executives of African start-up contest winners are White and foreign.[91] To wit, while Africapitalism's chief evangelist Tony Elumelu champions ethical leadership among corporate leaders, vested with substantial "private authority" in the neoliberal state, he has often sidestepped the democratic demands of Nigerians for state favor, most notably amid the #EndSARS campaign.[92] As Ochonu makes clear in the press, the moral leadership of another celebrated Africapitalist, Aliko Dangote (Africa's richest entrepreneur), has also been challenged by a track record of antiworker policies and cronyism.[93]

Typically, capitalist values are centered on the sustaining of growth, return on investment, market optimization, economic efficiency, and creation of stakeholder wealth, parameters that are often achieved through the monopolization of resources. As the critical economist Stefan Ouma states, "Non-shared value is not an aberration from an otherwise benevolent model of capitalism but fundamentally part of its history."[94] To equate these values with shared wealth and freedom is an artful turn of phrase. And in the discourse of Africapitalism, this false equivalency has led to the promotion of a naive ideology, perhaps, that African or indigenous-based forms of capital exchange will be more ethical simply because they are "African," that is, done by Africans. Entrepreneurism, here, is clearly cast as a social and political development strategy, and in many respects this Africapitalist aesthetic was shared by many of the activist developers I encountered in Ghana. But as I illustrate in the next section, the seams of such a neoliberal discourse at the state level became quite clear in Ghana in 2023–2024 with the rise of the #FixtheCountry movement and the authoritarian response.

#FIXYOURSELF: THE LIMITS
OF NEOLIBERAL NATIONALISM

Africapitalists, by emphasizing a Pan-African sensibility, are attempting to assert continental and racial autonomy in the space of business and international commerce. They are also advancing the laudable notion that an ethical relationship between markets, individuals, and governments could lead to greater self-sufficiency at both the local and international levels. However, the emergence of "Black diamonds" and Afropolitan elites has illustrated that such a focus on neoliberal growth alone has mixed results.[95] Though Africa's middle-class and consumer economy has grown since the new millennium, growth, equitable wealth and social prosperity for workers as a whole especially has lagged in proportion to overall population growth, with critics of Africa's export boom calling it flat or "jobless growth."[96] If Africapitalism at the macro level may yet have achieved such grand-scale results, it is certainly evident from both research literature and my experiences on the ground in Ghana that economic and social enterprises at the level of MSMEs has had an overall beneficial impact for everyday citizens including activist-developers, especially when paired with the practice of mobile money.

Shortly following my encounter with Kofi, I sought to understand the impact of mobile money use among entrepreneurs and everyday users, and I enlisted the help of a research assistant to conduct interviews with digital wallet users at key sites throughout Accra, including a mall, transit hubs, and mobile money reseller kiosks throughout the city. Person-to-person (p2p) digital wallet use in Ghana was just passing the early adoption stage, comprising nearly 30 percent of all financial transactions at the time.[97] From my own experience, though mobile wallet platforms had started as early as 2011, there were relatively few people I encountered up until 2016–2017 who were using them regularly. In interviews and surveys of more than 20 users,[98] what we found was that transactions through digital wallets allowed individuals to connect and maintain family ties and social obligations: parents of teenage and college-age children transferred small sums to their families to help with weekly expenditures; spouses sent money to partners for regular home purchases; dating partners sent digital cash as gifts; businessmen and women paid for supplies and paid employee wages; and informal sellers used WhatsApp and other social media to advertise small cottage businesses selling everything from sneakers to custom fashions. More serious traders were beginning to do exchanges for bulk cocoa butter and shea products, scrap metal sales, restaurant equipment, appliances, and electronics, all using mobile wallets. Couriers were also a key part of the ecosystem, facilitating exchanges from wholesalers, retailers/entrepreneurs, and customers. They themselves were also being paid in mobile money to facilitate the transactions. As a key practice in the informal economy, selling and trading via mobile money in tandem with social networking sites provided individuals second incomes in addition to their profession (their "main hustle" was a term often used). Increasingly users were paying for school fees, electricity bills, and other household services using these exchanges.

Most respondents to these queries talked about the tremendously time-saving nature of mobile wallets, including avoiding ATMs, which could not always be relied upon to have cash or be in service. The exchange of "short codes" for money (digital or physical) was described as a work-around to bank visits,[99] which many respondents said were difficult because of long lines or being inconveniently located. A health-products seller named Loraine whom we interviewed at the Accra Mall told us that mobile money was vital to her business. She advertised her products via social media and arranged for couriers to transport items to customers in the south of Ghana: "I'm in Kumasi, but I have customers in Accra. I send them products by bus and they send me my money using mobile money. I don't pass my money through people's hands, or I don't pass it through the banks which waste time with long queues. . . . When people buy my products, I don't see them. . . . Distance is not an issue. I can transact business and be confident I will receive my benefits."[100]

In the Makola market district, we interviewed a garment factory manager, Caroline, who also operated a storefront selling toys across town in her neighborhood of East Legon. Caroline said her toy business was beginning to become a profitable side business, which she attributed to the growth in her online customers. "I have customers paying me via MTN wallet. They see items on my Facebook or Instagram page and call to inquire. If I have the items, I send them the pictures using WhatsApp and we agree for them to send MTN mobile money. Afterwards, I get the courier to go and deliver to them. It's increased my business. This is because some customers don't have the time to travel around traffic to come to my store so because of mobile money they can do business with me."[101] Though Caroline had only been using the services for a few months, she had started to use mobile money for other important petty purchases as well. These included paying her home gardener, ordering fish from a local seller, getting pizza delivery, and paying for courier services for such purchases. She primarily used digital wallets through the telecom operator MTN, but also regularly paid money to users of other telecom companies, as well as bills for Internet and satellite TV, through the fintech smartphone app ExpressPay.

In general, the untaxed nature of mobile money was a boon to an economy that many had remarked for years offered few annual jobs in the formal sector, even for college graduates.[102] Digital transactions provided income opportunities in the informal sector and emerging MSMEs, though the transaction costs for doing so could often be significant, at 1 to 5 percent interest per transaction: These costs have varied over the years in response to competitive pressure among the mobile companies as well as the evolution of "momo" systems from USSD-based services (SMS short codes) to dedicated mobile apps like MPower Payments on smartphones, in the style of US-based companies like PayPal or Venmo. The impact was demonstrable on local economic growth and "financial inclusion" strategies, with use of formal banking due to mobile money partnerships rising as much as 30 percent in Ghana over the course of seven years.[103]

This fervor for entrepreneurship and growth in digital tools for liquid capital among everyday Ghanaians ran parallel to the overtly neoliberal sentiments of the New Patriotic Party (NPP), which became Ghana's ruling party with the election of Nana Akufo-Addo in 2016. Akufo-Addo, the son and nephew of Kwame Nkrumah's political rivals during the immediate postcolonial era, had often touted his party's business acumen prior to his election, which he claimed would inform policies helpful for economic development. The NPP government's central political identity was to offer a broad program of reforms based on government deregulation, decreased red tape, and facilitation of digital access in order to allow for more technological innovation and frictionless services. It became essentially the party of Africapitalism in Ghana. As Stephen Taylor noted, the NPP has approximated its policies and imagery to that of the conservative GOP in the United States and is an affiliate of the International Democracy Union, a global alliance of center-right parties.[104]

"Development for Freedom" has been NPP's central motto, and its signature campaign slogans over the years included the promise of "one village, one factory."[105] For many, in contrast to the previous NDC administration's lingering image as the caretaker of the postsocialist welfare state, the Akufo-Addo's government's economic policy could be best described as *let the market decide*. Akufo-Addo's government specifically projected state rhetoric eschewing "dead aid" and declared in several speeches Ghana's goal of becoming independent of global finance capital, especially funds from the IMF and World Bank. "We do not want to remain the beggars of the world, we do not want to be dependent on charity,"[106] Akufo-Addo said in the immediate aftermath of his hard-won reelection. In 2019 the administration inaugurated a campaign of "Ghana Beyond Aid," rejecting $1 billion in annual loans from the IMF. In short, Ghana during the years 2016–2024 embraced the rhetoric of a developmental state and in the process approximated the governance of nations such as Singapore and Rwanda, while still existing in an adversarial two-party system that has sustained its image as a model of democratic handovers since 1992.

While Ghana's profile has risen, most notably with its "Year of Return" heritage tourism campaign to diaspora Blacks in 2019, its economic challenges have mounted, spurred by increasing reliance on debt instruments such as the Eurobond and the crash in global supply chains during the 2020 coronavirus pandemic. The Akufo-Addo government had to back away from its anti-Western dependency rhetoric, accepting more than US$600 million in emergency loans in 2023, and after a year of wrangling with China and other creditors, US$3 billion in loans from the IMF in 2024. His administration subsequently raised income taxes, raised retail VAT by 2.5 percent, and during the pandemic imposed a 1 percent "recovery" tax on imports. It was the Akufo-Addo administration's imposition of an "e-levy" on mobile money transactions in 2022 (at first 1.5, then 1 percent of all exchanges over GHC100), however, that became a source of outrage, and furor erupted among Ghana's entrepreneurial class, emerging from both the pandemic and several years of rising energy prices and inflation. The GSMA has estimated that taxes on digital wallet transactions in particular caused a dramatic drop in the growth of mobile money in the nine months following its implementation, with transaction values and revenues falling up to 35 percent from the previous year.[107] A growing revolt of the middle class, particularly those running small enterprises, had been brewing in Ghana for several years, on the tail of nearly a decade of energy shortages. The resentment was captured in the enduring hashtag and political slogan #dumsor, the local term for electricity blackouts (translated as "lights-off-lights on"). Since 2015, #dumsor had been a rallying cry and point of criticism for the earlier Mahama NDC government, and in many ways had contributed to Akufo-Addo's eventual victory.

The year 2020, in addition to being the year the COVID-19 pandemic emerged, was also an election year, one in which Akufo-Addo's policies would take on an increasingly technocratic nature. A national ID system was implemented, with Ghanaians required to register their identity online and use a chip-encoded card to access public services. After a few years of failed attempts to get Ghana's famously engaged citizens to participate in mandatory SIM-card registries aimed at reducing fraud, the Parliament-approved law had the effect of limiting people's multi-SIM use and provided more transparent ways to monitor mobile wallet exchanges. As COVID-19 shut down travel and commerce, the costs of everyday materials began to rise in Ghana, exacerbated by the government levy on imports.[108] The strong executive decrees of Ghana's government seemed to lessen the pandemic's spread, shutting down international and regional travel through Kotoko International Airport and some public gatherings. But implementation of policies such Executive Instrument 63 (EI63)—the establishment of the Emergency Communications System Instrument—quickly began to stoke fears that the pandemic was simply for a pretense for increased government surveillance. As Smith Oduro-Marfo writes in "Transient Crises, Permanent Registries," the order left open ended what might be considered an "emergency," not limiting disturbances to health crises, and included no expiration clauses at the time. Amid sympathy strikes for #BlackLivesMatter and Nigeria's growing #EndSARS movement against police brutality, the Akufo-Addo government began to selectively enforce laws regarding public protests and civic demonstrations, especially during the election cycle. As Oduro-Marfo wrote, "In a country where security agencies are often at the beck and call of governmental actors, the arbitrary and self-serving use of such a surveillance system cannot be discounted."[109]

In 2021, from the frustrations over what were seen as politically expedient enforcement of COVID restrictions during the elections and a lack of government response to rising prices and infrastructure concerns emerged a grassroots hashtag campaign, petitioning for government accountability using the pointed phrase #FixtheCountry. In many ways #FixtheCountry was an evolution of the participatory slogan #GhanaDecides and the nihilistic cry of #dumsor. It quickly built off the momentum of previous Occupy movements in Ghana, including #OccupyFlagstaffHouse, another movement largely led by middle-class and social elites in 2016, decrying energy shortages in a country that had only a few years earlier begun

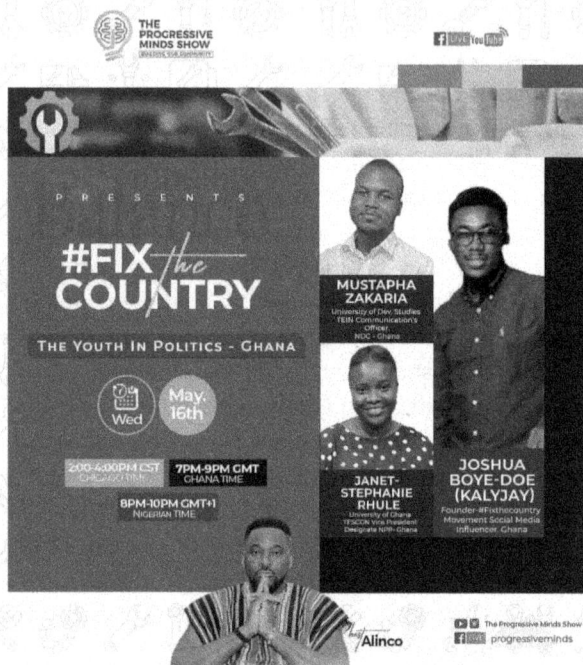

FIGURE 12. An online advertisement for an episode of *The Progressive Minds Show*, a Ghanaian and Black public affairs program hosted from Chicago. This episode featured the founder of the #FixtheCountry movement, comedian Joshua Boye-Doe, who uses the social media handle @Kalyjay. *Source: The Progressive Minds*, May 14, 2021.

exporting oil to the world.[110] Started by students and entertainers, #FixtheCountry, which criticized both of Ghana's leading parties, quickly drew mass appeal, and the hashtag went viral. Joshua Boye-Doe, a comedian and national service member at the time, was acknowledged as the creator of the hashtag, which he posted under his otherwise comedy-driven social media Twitter/X handle @Kalyjay. Boye-Doe, who had amassed more than 600,000 followers at the time, was interviewed on the Chicago-based *Progressive Minds* show, a diaspora-media video channel on Facebook and YouTube, a few weeks after the hashtag began trending. While he has been less active in the movement in recent years, his remarks captured the political potency of the meme at the time:

> [T]he fuel prices have gone up five times this year. Cement prices went from
> 40 cidis last year, and it's now almost at 50. So I feel the rate at which things
> are increasing is harsh on everyday citizens, especially if you're not seeing

results, because we pay taxes to get results. So, if I'm paying and I'm not getting results, I have every right to be worried And I feel this is what the government is missing. It's like imposing taxes is a way of fixing the economy right-o.
That's the laziest way of income, because for an economy to work, we need to
create more jobs, but they are rather imposing taxes on us. And if someone is
not working, they expect to pay tax and then it's going to put a burden on the
already working-class people that already pay tax. It is going to increase the
burden on them. This is why it's making the country very difficult.[111]

The early rhetoric of the movement was significant in that it signaled the seeming ways an Africapitalist ethos had pervaded public discourse. In the language of a neoliberal model in which "citizens" were
positioned as "customers" as a form of political entitlement, digital and
grassroots protest politics in the last decade in Ghana called out failures
of elected officials to address issues of national development endemic to
postcolonial ecologies of breakdown. The #FixtheCountry and earlier
#dumsor protest politics had signaled dissatisfaction with the govern-
ment for failing to facilitate infrastructural stability, and thus economic
prosperity, at a time when economic conditions in Ghana and Africa as
a whole were thought to be in the ascendant. In a swift and telling reaction to the mass movement, the hashtag #fixyourself quickly emerged,
with a prominent MP posting to Twitter/X: "Fix the Country! Gyimie!
('Stupidity!') You don't pay your taxes. You criticize every good policy.
You receive bribe[s]. You enjoy illegal connections. . . . Please fix yourself first!" The comment was taken down, and the official later apolo-
gized, advocating instead for "collective action and responsibility."[112] But
the sentiment flourished online, where #fixyourself in Ghana became a
retort to a movement largely portrayed as impertinent young people.
For Ghana media researchers Agana and Prempeh, #fixyourself "became
widely perceived as the establishment position in the debate."[113]

In a signal of increasing intolerance to the demands for civil reform,
a protest march in Accra scheduled for May 9, 2021, was shut down by
the Ghana Police Service under the authority of the COVID protections.
While a virtual protest and sympathy marches took place in diaspora cities like New York, the sense of repression began growing: Twitter/X soon
circulated images of a CitiFM journalist being arrested for covering the
controversy in Accra.[114] In a particularly chilling moment in June 2021, a
local civic watchdog using the social media handle Macho Kaaka ("Uncle
Strongman" in Hausa slang) was beaten to death by a mob for posting
critical videos online and using the hashtag #FixtheCountry. Kaaka, real
name Ibrahim Mohammed, was himself an NPP activist in the town of

Ejura in the Ashanti region, whose videos included exposés of trash-filled gutters on the streets of his hometown and other signs of neglect of public infrastructure by the local authorities. Among his most viewed posts on Facebook and WhatsApp is a long-take video highlighting a flooded sewer-way, filled with garbage and feces, cutting off access to homes. In the video a woman, whose only apparent access to her compound is via a small bamboo bridge over a large washed-out trench, is seen decrying the lack of government response. She shouts, "We have complained for so long and no any solution [*sic*]. But if it is election time you will come and fool us to vote for you. But after you get your votes, you don't mind us again."[115] Freedom House listed Mohammed's killing in its annual report on Ghana, in which it noted the country's increasingly downward ratings for journalism rights.[116] The death sparked days of protest in the region, and the police response to it was widely perceived as heavy-handed. Two protesters were killed by police gunfire, and more were critically injured. While a subsequent government investigation resulted in an apology and financial restitution to the victims,[117] many perceived the incident as a sign that civil liberties in Ghana were eroding, despite the rhetoric of expanding economic freedoms under the Akufo-Addo administration.

In 2023 the #FixtheCountry movement was rejuvenated by a series of protests in September, which were also met with government resistance. The movement was led this time not by a social media influencer and celebrity fanfare, but by a civil society group named Democracy Hub, led by lawyer and advocate Oliver Barker-Vormawor. The scheduled three-day protest included political pressure groups such as Arise Ghana, which also had been advocating against rising costs in Ghana and government corruption. Many protestors wore red and berets, a local sign of protest and anti-establishment resistance that also made direct connections to South Africa's Economic Freedom Fighters movement. These protests using the hashtag #FixtheCountry and #OccupyJulorbiHouse openly aimed at identifying economic inequality and rising costs in the country,[118] with again many of the protesters including entrepreneurs and activist-developers. Over the course of three days, dozens of protesters were arrested and held in detention when they approached the president's office and mansion, Jubilee House, though in most cases civil charges were never filed. Barker-Vormawor claimed that he was beaten in custody, while also being offered bribes to stop his political activities.[119]

The rise of the 2023 protests and crackdowns coincided with the emerging profile of an individual who would seem to embody the Africapitalist aesthetic. A series of billboards began to be spotted around major roadways in Ghana in late 2023, carrying the slogan "Leadership for the Next Generation. #TheNewForce." The billboards were striking, with digitally altered images of a straight-backed paternal figure, outfitted in either a tradition woven *batakari* or sharp Saville Row blazer. Distinctively, the man wore a traditional mask of unknown origin. Social media threads began to speculate that it was the work of real estate impresario Nana Kwame Bediako, a Ghanaian socialite and philanthropist who used the name "Cheddar" for many years as his public brand.[120] Bediako had been responsible for several high-profile real estate ventures in Ghana in recent years that brought luxury architecture projects to Accra's skyline, such as the Ritz Carlton, the Kwarleyz condominiums, and most prominently, No. 1 Oxford Street, an LED-covered building and car showcase that has given the dusty Osu district a Miami or Dubai appeal. Utilizing lavish and self-important public personas such as "Cheddar" and "Freedom Jacob Caesar," Bediako has gained attention in recent years for his outrageous fashion and bold statements. He won public favor by donating over 200 camping tents and water to residents impacted by spillage from an Akosombo Dam failure in 2023, via his aid group the New Africa Foundation. As part of his campaign, he offered to bring an ocean port to the city of Kumasi, situated 250 km from the coast.

Some speculated that Bediakos's viral marketing New Force push was in anticipation of the announcement of a presidential bid for the 2024 Ghana elections. Nevertheless, the foundation began to promote an event titled The Convention, featuring key contemporary Pan-Africanist activists from across the continent, scheduled in January 2024, at which the "man behind the mask" from the billboards would be unveiled. Using the slogan "Igniting the Voices of Africa," the event organizers described the meeting as the kick-off for a revitalizing of Pan-Africanist ideals, stating that there was "no better place on the continent" than Accra to begin this movement. Among the keynote speakers was Julius Malema, president of the Economic Freedom Fighters Party of South Africa, a firebrand politician and magnet for controversy for his anti-White rhetoric, anti-LGBTQ statements, and other political spectacles that resulted in his expulsion from the South African legislature. In addition, the event featured Peter Obi, Nigeria's failed 2023 presidential candidate, who drew strong support from youth activists, including #EndSARS supporters. Also included were P. L. O. Lumumba, a former director of Kenya's Anti-Corruption Commission and public intellectual who has been a leading voice for delinking African interests from the West in recent years, and Arikana Chihombori-Quao, a former director at the AU, who was fired in 2017 for accusing the French government of promoting "neocolonialism." Chihombori-Quao's most notable recent project has been the development of a Pan-African heritage and business park near Cape Coast called Wakanda City (see the conclusion).

Set to take place on January 7, 2024, at the nation's renowned Independence Square, the event was canceled the day before by a directive of the Akufo-Addo administration. An official from the Diaspora Affairs directorate announced that there had been an "unforeseen" state event that conflicted with the rally, and hundreds who had showed up were turned away by police tanks. The following day, at a press conference, sitting beside Bediako and the other orators, the former AU ambassador Chihombori-Quao stated, "So the message that we were going to deliver to our youth today was the message of hope, the message of resilience and the message that says the Africa we want can only be built by us for us. And for us to succeed at doing so we must be united that the African youth are Africa's future. Without them there is no future for Africa."[121]

In Ghana's increasingly entrepreneurial-centered social landscape, Bediako's Africapitalist persona and new Pan-African movement would have seemed to be a welcome development in the birthplace of Kwame Nkrumah, even if the young entrepreneur eschewed any political humilities: "I am nothing to be scared of. I came to you as your salvation," said Bediako after the cancellation. Akuffo-Addo's reaction may have been blatantly protectionist, despite the lack of political participation from the opposition NDC at The Convention. In many ways the event's cancellation demonstrates the limits of Africapitalism as a nonpolitical strategy to transform civil society through technology and private enterprise. With the stymying of participation, even from wealthy actors such as Bediako in Ghana's political life, the republic's political economy may be more likened to a state dancing between neoliberal democracy and market-bolstered authoritarianism, according to Edozie. This behavior is endemic among nations aspiring to become "developmental states." The developmental economist states, "Neoliberal authoritarianism represents a conflation of political authoritarianism and economic liberalism in opposition to democracy and especially in opposition to democratic constituent power."[122]

AFTER AFRICAPITALISM

In an article published amid Nana Akufo-Addo's first inauguration as president in 2016, cultural critic and Ghana scholar Jesse Weaver Shipley wrote that the NPP standard-bearer's election begged the question of "what political aesthetics tell us about future alternative economic configurations dressed in the attire of the free market."[123] Seven years after Ghanaians democratically endorsed a neoliberal state, it is now fitting to ask: What does Pan-Africanism dressed in the attire of the free market portend for the future of African liberation?

It may be unfair to equate the political opportunism of neoliberal statemen with Africapitalist idealism. Free enterprise, tech entrepreneurship, and disruptive innovation in finances have created new economic relationships for Africans in the 21st century, as seen in the scenarios with M-PESA, Flutterwave, and mobile money. Ultimately, it's important for readers to distinguish between an enterprise that advances social good and produces wealth and another form of enterprise that seeks to limit access to resources as a precondition of generating wealth. Africapitalism may simply describe ideas for an ethical ideology for management and economic systems and flows. But such a liberation ethos for entrepreneurs does not typically describe what we can think of as neoliberal capitalism; rather, it approximates ideas that are at their core socialist.

In many ways, activist-developers would embrace the seemingly libertarian economics of the Akufo-Addo administration, given the "anarchocapitalist" roots of hacktivism that drove Silicon Valley's start-up culture. What is interesting to note, therefore, is that while the NPP government advocated for neoliberal growth based on its ability to deliver economic and thus democratic rights to its constituents, its policies began to mimic the authoritarianism of African autocrats of the postindependence era. The Ghanaian government enthusiastically advocated for the AfCFTA using the Pan-African rhetoric of independence and anti-colonialism, claiming that free trade would produce not only economic freedoms but greater protections of human rights and freedom of movement. The Year of Return campaign in 2019 made great gestures akin to a Black internationalism, albeit one that openly courted diasporan capitalists, celebrities, and tourism as a tool of Ghana's development. Amid this impression of Africa Rising, Ghana took on a myriad of debt instruments, including Eurobonds, Chinese financing, and gold-for-oil schemes, but failed to halt inflation, ultimately prompting the country to accept neocolonial conditions once more. The costs of these maneuvers were transferred through taxes and levies, policies that ultimately have strangled MSMEs. In the case of the #FixtheCountry movement, the organized political resistance to these policies by small businesses and advocates was met with stern political reprisals using libertarian phraseology; more than partisan tribalism, #fixyourself communicated to advocates of reform and accountability that the problems of Ghana were not of the government's responsibility, but the burden of the neoliberal subject.

The people I have termed Pan-African futurists in this book are not necessarily supporters of a form of state planning that typified the Nkrumah and perhaps the Rawlings regimes; their actions are not necessarily tied to state policy and often lack a party affiliation, as in the case of #GhanaDecides. They do, however, act as insurgent system "disruptors," in that they draw on tactics of hacking, creative destruction/innovation. Despite their similar dedication to free enterprise principles, these actors ironically find themselves at odds with the state for its failure to secure social entitlements pertaining to personal and broad social wealth. Whereas Africapitalist's trickle-down virtues and allusions to "the African Dream" have proved patient if not reticent complements to the existing political order, activist-developers born of hacktivism and diasporic alienation have been integral to Ghana's hashtag movements and broader calls to transform Africa's intransient postcoloniality. As the ties of groups like GhanaThink reveal, activist-developers' connection to digital diaspora networks and their circulation abroad (virtual and physical) provides opportunities for the entrepreneurs' greater economic and political capital and to maintain sustainable livelihoods in Ghana.[124] Diaspora circulation for the activist-developers documented in this book has sustained not only livelihoods but a Pan-Africanist resolve that, while not necessarily socialist/internationalist in nature, is focused on a politics of social benevolence and social wealth in the homeland and for Africa in general.

CONCLUSION

The title of this last chapter was inspired by the writings of the 20th-century revolutionary Pan-Africanist and statesman George Padmore. Padmore was a key figure in the midcentury maturation of the Pan-Africanist movement. In his 1956 intellectual history and political memoir *Pan-Africanism or Communism*, Padmore vociferously asserted that a Soviet approach to Black liberation and Africa's development had no future. According to Padmore, the fledgling Communist International—a collection of state partyists in the Third World and local communist activists in Western metropoles—had failed to address the central question of race in world relations, perpetually asserting the primacy of class. Black liberation was used as window dressing for anti-Western propaganda in the period before and after World War II, Padmore asserted. He said, the USSR and other communist states provided few resources to address the plight of anti-Black racism or independence in the Caribbean. Adverse to the promises of the emerging development agenda supported by President Harry Truman and Western Europe, Padmore insisted that Pan-Africanism was the key collective ideology that could advance African independence and autonomy. In his seminal book, Padmore wrote: "Pan-Africanism looks above the narrow confines of class, race, tribe and religion. In other words, it wants equal opportunity for all. . . . Its vision stretches beyond the limited frontiers of the nation-state. Its perspective embraces the federation of regional self-governing countries and their ultimate amalgamation into a United States of Africa. . . . This is our vision of the Africa of Tomorrow-the goal of Pan-Africanism."[125]

Pan-Africanism or Communism, rather than being seen as a relic of mid-20th-century Black internationalist polemics, can better be understood as a prelude to the ideological struggles of Africans in the 21st century. For this author, this dilemma has endured in the more than 60 years since Ghana became the most prominent symbol of Black African independence. While Africapitalism remains a nascent or ameliorating philosophy, intended to assert some continental agency in what is increasingly described as a second "Gilded Age" of (digital) feudalism,[126] Africapitalists' celebration of a Pan-African ethos seems weakly historicized and politically disingenuous for the majority of mass movements and political actors who have struggled for Black liberation and African autonomy since the mid-19th century. Certainly in utilizing policy from the Soviet Union, Indonesia, or China, emerging out of state planning, autocratic socialist regimes in Africa provided many failed experiments and fewer triumphs in the 1970s. But I argue that while these projects are normally criticized for their authoritarian political tendencies and the lack of experience of their administrators, the newly independent governments of Africa and the Black Caribbean diaspora were primarily concerned with building economic independence, in the very same ways that the Africapitalists aspire to do. As Marxists, they routinely dismissed the equation of capitalism with freedom and broader notions of social wealth.[127] Though Africapitalism as a distinctive or alternative model of capitalism may consist of a unique praxis and perhaps indigenous sensibility around *ubuntu*, its vision of Pan-African political sovereignty remains elusive. It is unlikely that embracing such radical forms of digital capital such as blockchain or crypto-exchanges will alter these terms.[128]

A central underlying thesis here is that technology cannot be depoliticized. It may be ambivalent to its uses, but its political effects are socially determining. All structures are not configured equally. Not all users are equals. The futurist dimensions of Pan-Africanism cannot be predicated on the belief that an unequal market and lopsided form of globalization can be made fair by the adoption of technologies from the West, which are largely responsible for these inequalities. Pan-African futurism, then, must be more Pan-Africanist than futurist, as political as it is digital. As digitality and global capitalism have allowed for greater movement of wealth, ideas, and goods, we might laud the fact that an Africapitalist vision would potentially also allow for the expansion of humanist freedoms. But a futurist Pan-African disposition must be focused on an understanding of collective and shared wealth and prosperity that is broader than the praxis of a seemingly unregulated "free market."

Padmore's enduring query posits the paradox between Africapitalism and Pan-Africanism or a Pan-African futurism and proposes a fundamental question for activist-developers at home and abroad. As economies like Ghana continue to utilize ICT to become more stable and potentially structurally responsive to their citizens, will future African polities be markets that advance wealth based on socially benevolent "African values," or simply advance the existing ethos of neoliberal markets grounded in scarcity and scale with new geographic agents? Will the zones of opportunity created by homeland-diaspora enterprises propel this developing nation toward a Wakanda-like utopia, or will a "pan-African" expatriate capitalism come to typify its internationalist focus?

Conclusion

The Aesthetics and Technopolitics
of Pan-African Futurism

In this book I have explored the promise of technology and innovation as liberatory practices, in the rhetoric of Ghanaian activist-developers, and in diaspora media. These projects have included the very human-centered practice of gaining access to ICT via the (re)circulation of mobile phones, diaspora-to-homeland websites, and other inventive digital connection work-arounds such as MagicJack and multi-SIM-card use. This also includes traditional forms of innovation such as app and service building that defy enduring notions of Africa's anti-modernity and challenge concepts of technological diffusion being primarily from the West to the Global South. Using these tools, Ghanaians are hacking development, in the parlance of Silicon Valley—innovating from below and producing new material techniques through their design, strategic use, and redeployments of technology; They have appropriated ICT-infused concepts, such as hacking and social entrepreneurship, and localized Silicon Valley networking events such as BarCamps, hackathons, and other skills-building events throughout the homeland. They have utilized web-based social media to promote civic accountability campaigns, and filmmakers and musicians have seized upon the viral potential of YouTube content, often using the imagery or spark of Afrofuturism and Africanfuturism. Community events such as GhanaFest in Chicago and the Ghana Tech Summit in Accra have virtualized their gatherings, producing digital diasporas, creating virtual third spaces for Ghanaian national identity prior to the COVID-19 pandemic and in the times since. These spaces of digital discourse, including GhanaWeb message boards, WhatsApp threads, Facebook groups, and online broadcast sites such as *The Progressive Minds Show*, make indistinguishable the polities of home and abroad, as Ghana (and Africa) remains at the center of their discourse and social imaginary. Mobile money and digital finance have been explored as a means of financing personal autonomy and seen as a means of growing the wealth of the nation, in spite of both global tactics of financial redlining, government overreach, and the itinerant sociotechnical systems of postcolonial states.

The COVID-19 pandemic and its political-economic wake threw techno-optimistic projections of Africa's digital uptake into a quagmire of varying outcomes. According to the ITU, Internet and mobile phone usage dropped dramatically across the continent during the pandemic, and since then it has recovered unevenly.[1] Yet the growth in fiber-optic sea cables, such as Google's Equiano line in 2022 and China and Meta's 2Africa line in 2024, has had a dramatic impact on connectivity. "Middle-income" African nations with Internet penetration hovering for a full decade around 40–60 percent are beginning to seeing nationwide usage rise to as much as 90 percent in urban areas. Still, there remains much to be done: Most of the nations at the bottom 20 of the UN Internet development index (IDI) are located on the African continent, with Uganda, Central African Republic, and Eritrea having penetration rates close to 10 percent, and otherwise economically bustling states such as Ethiopia having mobile phone usage less than 50 percent (national averages).[2]

Against the enduring asymmetries of the networked world, the strategies used by Ghanaians in this book advance a practice of digital inclusion, using ICT and social entrepreneurship for autonomy, and often are Pan-African in character, reflecting the transnational political legacy of the republic since the Nkrumah years. In contrast to the technologies of abjection typically associated with African digital entanglements as discussed at the outset of this book (e.g., coltan extraction, *sakawa* or 419 email fraud, e-waste), I believe the following approaches characterize the broad set of socially benevolent, Pan-Africanist aesthetics utilized by Ghanaians and other Africans during the years of my research:

- *Disruptive innovation*: ICT has been deployed to "hack" social problems, especially long-enduring infrastructure issues of breakdown and redlining. Pan-African futurists work as disruptors to engage with issues of access to Internet, telephony, and ICT tools themselves, such as mobile devices and computers. They have also

FIGURE 13. A film editor works on a project at Impact Hub Accra under an image of Kwame Nkrumah in 2016. Photo by author.

reappropriated tools for financial exchange, refashioned communicative platforms such as WhatsApp into income-generating media, and transformed mundane digital tools into a means for building community with dispersed relations abroad, recalibrating technology amid experiences of sociotechnical experiences of exclusion (e.g., MagicJack, SMS, TTY, Hangouts).

- *Technological accessibility:* Activist-developers and digital diasporans have deployed technology that is informal, decentralized, and increasingly reliant on digital orality and affective media, advancing African ways of knowing, learning and interacting. Mobile devices (especially low-tech feature phones) are the primary African digital media platform. In the coming years, this is likely to include voice-driven technology, local-language user services, voice-driven interfaces on mobile apps, and media cultures continuing to center Internet radio and podcasting. Other tactics of access include innovation with tools highly relevant to the user base and user needs, utilizing low-cost, austere bandwidth and computing power, typified by practices such as *tethering*, multi-SIM use, and applications that utilize text messaging as a primary form of communication and financial access, such as "mobile money."

- *Democratic agency:* A sense of political horizontality is built into tool use and iterative design in Ghana and Africa broadly, advancing the desire for accessibility already ascendant in techno-utopian concepts such as the network society. Examples include the #GhanaDecides campaign; Odekro, the parliamentary policy tracker service and advocacy group; and skills-building network events such as BarCamps, utilized by groups such as GhanaThink and DevCongress, created to generate greater accessibility to ICT, coding, and entrepreneurship opportunities. Through the practices demonstrated, such as online meetups via video chat (Google+Hangouts) and collaborations emerging from coworking spaces such as Impact Hub Accra or iSpace, ICT designed among activist-developers during this period were generally aimed at mass usage in Ghana and geared toward greater connections and social and political engagement.

These three categories of practice have collectively characterized the prosocial aspects of locally constructed ICT in Ghana and its digital diasporas and reflect a guiding set of aesthetics for African digital media culture since the early 2000s.[3] These are in, short, the aesthetic ideals of Pan-African futurism, which exemplify political commitments to the vision of an autonomous Africa, utilizing media and information technology to assist in the continent and its people's social, economic, and political transformation in our times. These aesthetic ideals in many ways attempt to stabilize the flexible technological affordances of ICT within a Pan-Africanist paradigm. As an ethnographic project, this book has provided both an interpretation of these techniques and a useful set of labels for these interrelated stories during the Internet's expansion in Africa between the 2009 and 2024.

I end this volume with a description of a few late-developing social projects that demonstrate the Pan-African futurist spirit: a commitment to sociotechnical innovation and collectivist notions of social wealth and autonomy.

WAKANDA AS PRAXIS

Beginning in 2018, as the Ryan Coogler/Marvel films *Black Panther* and its sequel *Wakanda Forever* began to become a global sensation, with some of the highest movie attendance rates both pre- and postpandemic, I became inspired and intrigued by the rise in discourse about Afro- and Africanfuturism that followed, given the movie's science fiction interventions and its techno-vernacular themes. I was also discouraged that attention was less often given to progenitors of the Afrofuturist aesthetic, including Sun Ra in the United States; Keith Piper, Kwado Eshun, and the Digital Diaspora film group in the UK; and Kenya's filmmaker Wanuri Kahiu. A more irksome experience, however, was that invariably after I would give a lecture on my research with activist-developers and Ghana's digital diasporas, an attendee would remark to me, "All this tech innovation in Africa sounds very exciting. You must be talking about Wakanda." Rather than lament this reductivism, I have taken it as a point of critical departure, and certainly American audiences are not the only ones who have made this connection. In *What Do Science, Technology and Innovation mean for Africa?*, MIT professor Chakanetsa Mavhunga, a historian of ethnoscience on the African continent, states: "The concepts of [science technology and innovation] matter at this specific historical moment in Africa, because there seems to be a feeling that Africa's time has come. This, Africa is rising narrative, is all over the World Wide Web, often under the name Afrofuturism."[4] The Zimbabwean-born professor is not alone in connecting techno-optimism about Africa with Afro- and Africanfuturism. As I have stated at the outset of this project, Afrofuturism and Pan-African futurism are not necessarily competing ideas: many technology innovators on the continent and in the diasporas have drawn tremendous inspiration for their work from the *Black Panther* series and the African futurist film *Pumzi*, as well as the novels of Octavia Butler or Nigerian author Nnedi Okrafor, and from mainstream (mostly White) science fiction writers, films and video games.

In my view, Pan-African futurism as a concept situates the practices and site of innovation in STEM in African-led efforts that are currently active, on the ground, and taking place right now. These practices are both forward looking and reflective of the past, thus abandoning a teleological sense of time and development. African and diasporic sensibilities of time are crucial to this aesthetic, departing from both the Hegelian notion of an Africa without a past and the neoliberal sense of history as having already unfolded. A sense of the Akan notion of *sankofa*[5] is rather embedded in both the technology and discourse in projects such as Leti Arts, a video game company started by Wesley Kirinya, based in Nairobi, and Eyram Tawia, a developer and designer I have interviewed several times since 2011. Their titles, such as "African Legends," feature superhero takes on classic characters of African folklore, such as Ananse the spider, who is portrayed as a technologically savvy trickster.[6]

The 2019 Afrochella music festival rebranded itself Afrofuture in 2022. The London Business School's 2023 African Business conference used the slogan "The Future Is African," while the California-based African Diaspora Investment Symposium recently touted itself as "Builders of Africa's Futures." But there is perhaps currently no effort more illustrative of the connection between Pan-Africanism and African futurism than the project to create a tech innovation hub near the site of the genesis of the Transatlantic Slave Trade in Ghana—an effort called Wakanda One, the City of the Return. In 2021 Dr. Arikana Chihombori-Quao launched a bold initiative centered in Cape Coast, Ghana, calling her project Wakanda City—an obvious nod to the fictional city-state at the center of the Afrofuturist movie epic *Black Panther*. In the films and long-running comic book series, Wakanda is a peaceful and self-sufficient kingdom centered in an otherwise turbulent Africa. The source of its power is the nation's exclusive province of a fictional mineral resource ("vibranium"), which powers industry, commerce, and the military, in technological and ecological harmony. Chihombario-Quao's own organization, the African Diaspora Development Institute (ADDI), seems like a civil society NGO lifted from the movie itself, as its mission states:

> We have come to the realization that the current status of the African economy is due to the mass exodus of the African Diaspora out of Africa. The exodus began with the loss of Africa's children through slavery 400 years ago,[7] followed by the recent migrants who left Africa in search of greener pastures or running away from famine and wars. The children of Africa have ended up in the Americas, Australia and New Zealand, Europe, and Asia.

African Heads of State are calling on all people of African Descent (African Diaspora) to come back home and build the Africa we want.[8]

In its promotional literature, the city Wakanda One is envisioned as an ultramodern, port and land development project, which includes designs for a tourism complex, residential homes, office spaces, and a trade exposition. The project plans to open in Cape Coast, the city 140 kilometers west of the capital Accra, which served as the Gold Coast region's most active slave-trading port and later as the capital for the what would become Britain's Crown colony during the 19th century.

Originally from Zimbabwe, Dr. Chihombori-Quao worked for many years as a physician and executive in the United States, before becoming a leader in the AU's public health directorate. She then served as permanent AU ambassador to the United States in the mid-2010s. In 2019 Chihombori-Quao's tenure in that post was terminated following public statements she made criticizing the country France for what she termed "continuing colonialism" on the continent—a firing that sparked public outrage and an online petition by her supporters. But the attention allowed her to gather momentum for her Wakanda One project and the work of the ADDI. In a video launch event in 2021 on YouTube, Dr. Chihombori-Quao stated: "Our flagship program is going to be the Wakanda City of Return. . . . That is going to be the shining city on the hill, the city that's going to stand up and look across the Atlantic and remind the slavers that while you may have won for over 400 years, the children of Africa are back home . . . the children of Africa, descendants of the formerly enslaved are going to converge and look the slavers in the face and say . . . we are the victors." The Wakanda project exists mostly on paper at this time. But momentum for a Pan-African future return has been building for years. In 2019 Nana Akuffo Addo's administration commenced the "Year of the Return" tourism campaign, enticing Ghanaian expatriates and the Black diaspora community worldwide to seek refuge in Ghana, especially in the years following increased racial violence in the United States and elsewhere.

A Pan-African futurist moment, referenced at the beginning of this book, also strikes a similar accord. In 2020 the annual Ghana Tech Summit was entering its third iteration. The global COVID-19 pandemic had pushed all events back several months in Ghana, and this year the conveners, like Chicago's GhanaFest organizers, opted for a virtual summit, with nine hours of broadcast talks and meetup sessions, many of them streamed live. The Ghana Tech Summit, founded by millennials, husband

and wife Einstein and Christine Ntim, kicked off a few years earlier in Accra, following the example of other Ghanaian networking and developer meetups in the preceding decade, such as BarCamp Accra. The son of a former Ghana communications minister, Einstein Ntim was born in Ghana and raised in the UK, where he would play professional rugby and later serve in the armed forces. Einstein received a degree from the London School of Economics, worked in the corporate sector for firms such as UBS and Deutsche Bank, and began several health-care companies. Eventually, Ntim would go on to get training and sit on the board of the Singularity University, a techno-utopian academy started by inventor and artificial intelligence futurist Ray Kurzweil. The Ntims duo had been active for several years, hosting "business accelerator" events in Haiti, where Christine's family has roots, operating under the banner the Global Startup Exchange. By 2020 GSE events included an annual "African Futures" business summit, as well as the Ghana Tech Summit (GTS).

Whereas BarCamps and Diaspora Camp conferences used their efforts to network with and educate aspiring youth and budding tech professionals primarily operating under the auspices of civil society and not-for-profit organizations, GTS is a business and set of conferences that help software firms such as SAP and management companies such as Deloitte make in-roads in Ghana's emerging developer community, which was sufficiently established and winning international awards by 2018. Despite extensive experience with Fortune 500 companies in the US and in Europe, the pandemic-era GTS conference struck a decidedly Black liberationist tone, as evidenced in Einstein Ntim's opening remarks. Broadcast via YouTube, Ntim spoke over scrolling web banners from sponsors such as LinkedIn, Amazon, Google Cloud, Microsoft, and Zendesk. In the speech Einstein waxed prophetic, stating that using information technology to develop Africa was essential to prepare for the "Fourth Industrial Revolution"[9]—and that in the immediate aftermath of the pandemic, artificial intelligence and globalization would bring more change to the lives of Africans in 10 years than had been experienced in the past 100 years. In words reminiscent of a famous speech in which Nkrumah envisioned "cities of Ghana becoming the metropolis of science, art, industry,"[10] the business developer Ntim stated:

> Africa's future is a singularity.[11] It's in a situation where we can rise to the challenge of this rapid acceleration of technology and we can meet the task and build the cities of tomorrow that we can be proud of that will be self-sufficient and serve our people effectively. We can use this time period that the pandemic has caused for us as an incubation period to learn about tech

and digital transformation, so that we can empower communities whether the person is in Bolgatanga, Tamale, or Takoradi so that they can still do effective work, get paid and advance their societies and communities. . . . [T]hrough technology, Africa can truly leapfrog this innovation curve that we're going through. I believe that Ghana can be the Black star, the champion to move the needle further forward.[12]

At another moment in the virtual plenary, Derek Addo, a GTS coordinator, echoed Ntim, focusing on the role of culture in advancing a Pan-African futurism:

Whether it's your name or where you come from, I'm Asante, I am descended from kings and queens. When you understand your heritage and the gravity of where you come from, just how beautiful and rich our culture is, I couldn't be prouder. That's one of the main things we take for granted and at the same time, it's right in front of us. . . . There's the shared creativity of our people. If you can make a business in light of [our circumstances] you're doing something special. I feel like we are the kings of doing more with so much less than the Western world have. Our shared creativity and ingenuity is boundless.

The sentiment was echoed by another panelist, Christina Glancy, an African American from Baltimore, who stated, "I am inspired, Ghanaians are so rooted, so they are looking for something else." Arnold Sarfo-Kantaka, a fellow management consultant who has worked with GTS and built his own research-based diaspora network for Ghanaians working primarily in Europe, stated, "What we're trying to build, is this ecosystem that can actually be a trusted incubator that can support job creation."[13]

Finally, the pandemic has also spawned digital social imaginaries around Afropolitanism and the Afropolitan mediascape that were barely foreseen at the start of this project. Amid the grimness of global shutdowns and stay-at-home orders, two Nigerian Americans—Eche Emole, a former law school student and digital marketer, and Chika Uwazie, a tech founder supported by 500 Startups—developed their own virtual community to establish camaraderie and share news on the audio-driven social media app Clubhouse. Garnering some of the platform's largest listenerships (more than 140,000 in 2022), the pair have since leveraged their audience into a nascent "digital nation" calling itself Afropolitan. Working with their own membership-driven platform (Afropolitan .io), a cryptocurrency, and having issued over 500 NFTs,[14] the goal of this start-up is to create a "minimum viable network state," open primarily to citizenship for Africans on the continent and Black diasporans worldwide. Currently negotiating the purchase of land in Zambia, they also plan to develop their own internal economy, which will operate in Afropolitan diaspora enclaves ("charter cities") in other sites across the globe, much in the way embassies do. In an interview with CoinDesk, a well-known cryptocurrency news outlet, Uwazie claimed to have raised over US$2.1 million for the project. Rather than averring to an anarcho-capitalist ethos for the Afropolitan project, Uwazie offered instead Pan-African commitment: "What's happening is that we're actually hoping to act as state actors. So, a great example would be in Ukraine. A lot of Africans were stuck in Ukraine [at the start of the war], a lot of students schooling there. If we actually have a community, we have a treasury, we have working groups, we could have stepped-in and helped on behalf of some of these African countries."[15]

THE TECHNOPOLITICAL COMMITMENTS OF PAN-AFRICAN FUTURISM

Inherent in the three aesthetics of a Pan-African futurism that I have discussed earlier and sentiments expressed previously by Ghanaians and Africans from around the continent are technical commitments to ICT for development, ethical obligations of expanding economic inclusion, and the progressive goals of sustainable development. They are better understood as a set of impact-oriented or action/research developmental approaches, that in the African and diasporic context incorporate decolonial and anti-racist sentiments. If ultimately the goal of tech enterprises, policies, and social movements incorporating ICT is to be focused on promoting social good or democratic agency in the Black world specifically, it is clear that these are informed not simply by nation-based patriotism, but by a set of unifying Pan-Africanist commitments. These I have already identified as the Pan-African sensibilities in the developer communities and digital diasporas during the research period for this book and in roughly the last 20 years of digital media innovation in Ghana. The sense of optimism and the commitment to material and humanist-oriented change embedded in these efforts are fundamentally futurist in orientation. The concept of Pan-African futurism can ultimately be used to characterize those who are striving for such transformation. This produces a particularistic material discourse that moves beyond an agnostic description of the affordances of technology and the "apolitical" dispositions of activist-developers and civil society activists. As the successes and drawbacks of #GhanaDecides, #dumsor, and #FixtheCountry illustrate, technological agency requires a politics, whether or not the aims of those digital movements are expressly political. The goals of development are inherently political, not merely instrumental, especially if actors seek to transform the existing world system of "unequal development/exchange." This is not to say that Pan-African interests need necessarily be pro-West or pro-China, or necessarily lean toward the strategic alignments with BRICS (Brazil, Russia, India, China, South Africa, Egypt, Ethiopia, Indonesia, Iran, and the United Arab Emirates) in the Middle East or Russia. Rather, Pan-African futurism, again, is centered in Pan-African solidarity and coordination. Africa's project of self-reconstruction and autonomy will inherently be seen as a political project, centered on negotiating the distribution of resources toward philosophies such as *ubuntu,* as the edges of the network society are brought closer together. Futurity need not be speculative, if innovation is already underway.

Notes

INTRODUCTION

1. Kwabena Agyare Yeboah, "We Need to Talk About Ghana's Year of Return and Its Politics of Exclusion," *African Arguments* (blog), December 19, 2019, https://africanarguments.org/2019/12/ghana-year-of-return-politics-of -exclusion/.

2. Graphic Online, "Akufo-Addo Grants Citizenship to 126 Africans in the Diaspora," *Graphic Online* (blog), November 28, 2019, https://www.graphic .com.gh/news/general-news/akufo-addo-grants-citizenship-to-126-africans-in -the-diaspora.html.

3. Visit Ghana, "Year of Return: 126 African Diasporans Granted Ghanaian Citizenship," December 19, 2019, YouTube, https://www.youtube.com/watch?v =7PK8cy4fdoo.

4. Ato Ulzen-Appiah, presentation at Ghana Tech Summit, Accra International Conference Centre, December 13–15, 2019.

5. Jan van Dijk, *The Digital Divide* (Polity, 2020); Alliance for Affordable Internet, "Advancing Meaningful Connectivity: Towards Active & Participatory Digital Societies," May 24, 2022, https://a4ai.org/research-database/.

6. Groupe Spécial Mobile Association (GSMA), *The State of Mobile Internet Connectivity 2019*, 2019, https://www.gsma.com/mobilefordevelopment /wp-content/uploads/2019/07/GSMA-State-of-Mobile-Internet-Connectivity -Report-2019.pdf.

7. Tania Begazo et al., *Digital Africa: Technological Transformation for Jobs* (World Bank Group, 2023). Internet usage and access divides are pervasive in the country of Ghana, with Internet access as high as 70–90 percent in major cities and towns and as low as 2–25 percent penetration in some rural regions. This gap persists also between the wealthier southern/coastal regions of the

country compared with regions north of Kumasi. The gap between men and women users is about 10 percent.

8. Hakim Adi, *Pan-Africanism: A History* (Bloomsbury Academic, 2018), 2.

9. Samuel Fury Childs Daly, "Ghana Must Go: Nativism and the Politics of Expulsion in West Africa, 1969–1985," *Past & Present* 259, no. 1 (May 2023): 229–61, https://doi.org/10.1093/pastj/gtac006.

10. Kwame Nkrumah, "African Socialism Revisited," in *Africa: National and Social Revolution*, At-Talia and Problems of Peace and Socialism Conference (Peace and Socialism Publishers, 1967), https://www.marxists.org/subject/africa/nkrumah/1967/african-socialism-revisited.htm.

11. "Africa Rising," *The Economist*, December 3, 2011, https://www.economist.com/leaders/2011/12/03/africa-rising.

12. Madanmohan Rao, *Mobile Africa Report 2011: Regional Hubs of Excellence*, Mobile Monday, November 1, 2011, http://www.mobilemonday.net/reports/MobileAfrica_2011.pdf; GSMA, "Sub-Saharan Africa Mobile Economy 2013," *Groupe Speciale Mobile* (blog), November 1, 2011, http://www.gsma.com/mobileeconomyssa.

13. Erik Hersman, "Mobilizing Tech Entrepreneurs in Africa (Innovations Case Narrative: iHub)," *Innovations: Technology, Governance, Globalization* 7, no. 4 (October 1, 2012): 59–67, https://doi.org/10.1162/INOV_a_00152.

14. World Bank, *Ghana: World Development Indicators: 2014*, 2014, https://elibrary.worldbank.org/doi/abs/10.1596/978-1-4648-0163-1; US Department of State, "Background Notes: Ghana Economy," accessed September 23, 2011, https://www.state.gov/u-s-relations-with-ghana/#history.

15. World Bank, *Ghana*; see also Henning Melber, ed., *The Rise of Africa's Middle Class: Myths, Realities and Critical Engagements*, Africa Now (Zed Books, 2016).

16. Iginio Gagliardone, *China, Africa, and the Future of the Internet* (Zed Books, 2019).

17. In 2019, Jack Dorsey's Twitter also announced that Ghana would be home to its African operations. These offices closed in November 2022.

18. Bruno Latour, "Technology Is Society Made Durable." *The Sociological Review* 38, no. 1, (May 1, 1990): 103–31.

19. Adi, *Pan-Africanism*.

20. The Caribbean country Haiti arguably became the first Pan-African nation-state in 1807.

21. Alvin Toffler, ed., *The Futurists* (Random House, 1972).

22. Anne Bowler, "Politics as Art: Italian Futurism and Fascism." *Theory and Society* (1991): 763–94; George L. Mosse, "The Political Culture of Italian Futurism: A General Perspective," *Journal of Contemporary History* 25, no. 2 (1990): 253–68. Bowler describes Italian futurism as a movement focused on "the sublation of art into a new revolutionary praxis that would transform the organizational fabric of everyday life" (Bowler, "Politics as Art," 764).

23. Mark Dery, "Black to the Future: Interviews with Samuel R. Delany, Greg Tate, and Tricia Rose," in *Flame War: The Discourse of Cyberculture*, ed. Mark Dery (Duke University Press, 1994), 179–222; Kodwo Eshun, "Further

Considerations on Afrofuturism," *CR: The New Centennial Review* 3, no. 2 (2003): 287–302.

24. Nnedi Okorafor, "Africanfuturism Defined," *Wahala Zone* (blog), October 19, 2019, http://nnedi.blogspot.com/2019/10/africanfuturism-defined.html. Okorafor uses the signature spelling "africanfuturism" to treat the concept as one distinct term.

25. Nnamdi Azikiwe, *The Future of Pan-Africanism* (London: Nigerian High Commission, 1961); Kwame Nkrumah, *Towards Colonial Freedom: Africa in the Struggle Against World Imperialism* (Panaf, 1973); Julius K. Nyerere, "A United States of Africa," *The Journal of Modern African Studies* 1, no. 1 (March 1963): 1–6.

26. Reiland Rabaka, ed., *Routledge Handbook of Pan-Africanism* (Routledge, 2020), https://doi.org/10.4324/9780429020193.

27. Reynaldo Anderson and Charles E. Jones, eds., *Afrofuturism 2.0: The Rise of Astro-Blackness* (Lexington Books, 2016).

28. Francis B. Nyamnjoh, *#RhodesMustFall: Nibbling at Resilient Colonialism in South Africa* (Langaa Research & Publishing CIG, 2016).

29. See African Diaspora Development Initiative, "Wakanda One: The City of Return," accessed December 21, 2021, https://ouraddi.org/wakanda-one/. More in this book's conclusion.

30. Okoth Fred Mudhai, *Civic Engagement, Digital Networks, and Political Reform in Africa* (Palgrave-MacMillan Press, 2013); Mohamed Zayani, *Networked Publics and Digital Contention: The Politics of Everyday Life in Tunisia* (Oxford University Press, 2015); Charles Villa-Vicencio et al., *The African Renaissance and the Afro-Arab Spring: A Season of Rebirth?* (Georgetown University Press, 2015).

31. Reginold A. Royston, "Reassembling Ghana: Diaspora and Innovation in the African Mediascape" (PhD diss., University of California, Berkeley, 2014); Safiya Umoja Noble, *Algorithms of Oppression: How Search Engines Reinforce Racism* (New York University Press, 2018).

32. Nicolás Valenzuela-Levi, "The Written and Unwritten Rules of Internet Exclusion: Inequality, Institutions and Network Disadvantage in Cities of the Global South," *Communication & Society* 24, no. 11 (August 18, 2021): 1568–85.

33. Julius Endert, "Despite Ghana's Commitment to Internet Expansion, Problems Persist," *DW Akademie* (blog), November 29, 2018, https://akademie .dw.com/en/despite-ghanas-commitment-to-internet-expansion-problems -persist/a-46508524.

34. Telegeography, *The State of the Network* (Telegeography, 2022), https:// www2.telegeography.com/download-state-of-the-network..

35. Manuel Castells, *The Rise of the Network Society* (Blackwell Publishers, 1996).

36. Terence K. Hopkins and Immanuel Maurice Wallerstein, *World-Systems Analysis: Theory and Methodology*, Explorations in the World-Economy (Sage Publications, 1982); Brian Larkin, *Signal and Noise: Media, Infrastructure, and Urban Culture in Nigeria* (Duke University Press, 2008).

37. Rita Raley, *Tactical Media* (University of Minnesota Press, 2009).

38. James Ferguson, *Global Shadows: Africa in the Neoliberal World Order* (Duke University Press, 2006).

39. Jenna Burrell, *Invisible Users: Youth in the Internet Cafés of Urban Ghana* (MIT Press, 2012).

40. Thierno Thiam and Gilbert Rochon, *Sustainability, Emerging Technologies, and Pan-Africanism* (Palgrave Macmillan, 2020), https://doi.org/10.1007/978-3-030-22180-5; Clapperton Chakanetsa Mavhunga, ed., *What Do Science, Technology, and Innovation Mean from Africa?* (MIT Press, 2017); Bruce Mutsvairo ed., *Digital Activism in the Social Media Era: Critical Reflections on Emerging Trends in Sub-Saharan Africa* (Springer, 2016); Bitange Ndemo and Tim Weiss, eds., *Digital Kenya: An Entrepreneurial Revolution in the Making* (Springer, 2016); Seyram Avle, "Articulating and Enacting Development: Skilled Returnees in Ghana's ICT Industry," *Information Technologies and International Development* 10 (December 2014): 1–13.; Hopeton S. Dunn et al., eds., *Re-Imagining Communication in Africa and the Caribbean: Global South Issues in Media, Culture and Technology* (Palgrave Macmillan, 2021).

41. Alphonce Shihundu, "There Are Many Ways You Can Be Scammed on the Web in Africa," *Dandc* (blog), June 28, 2023, https://www.dandc.eu/en/article/digital-fraud-major-challenge-africa; Victor Oluwole, "Top 10 African Countries with the Most Electronic Waste," August 29, 2023, https://africa.businessinsider.com/local/lifestyle/top-10-african-countries-with-the-most-electronic-waste/8czptln; Emma Woollacott, "Russian Trolls Outsource Disinformation Campaigns to Africa," *Forbes* (blog), March 13, 2020, https://www.forbes.com/sites/emmawoollacott/2020/03/13/russian-trolls-outsource-disinformation-campaigns-to-to-africa/; Oluwole Ojewale, "What Coltan Mining in the DRC Costs People and the Environment," *The Conversation* (blog), May 29, 2022, https://theconversation.com/what-coltan-mining-in-the-drc-costs-people-and-the-environment-183159.

42. James Yékú, *Cultural Netizenship: Social Media, Popular Culture, and Performance in Nigeria* (Indiana University Press, 2022); Msia Kibona Clark et al., eds., *Pan African Spaces: Essays on Black Transnationalism* (Lexington Books, 2019); Jonathan Donner, *After Access: Inclusion, Development, and a More Mobile Internet* (MIT Press, 2015); Burrell, *Invisible Users;* Abou B. Bamba, *African Miracle, African Mirage: Transnational Politics and the Paradox of Modernization in Ivory Coast* (Ohio University Press, 2016); Stephan Miescher, *A Dam for Africa: Akosombo Stories from Ghana* (Indiana University Press, 2022); Abena Dove Osseo-Asare, *Bitter Roots: The Search for Healing Plants in Africa* (The University of Chicago Press, 2014); Mavhunga, *What Do Science, Technology?*; Abdul K. Bangura, *African Mathematics: From Bones to Computers* (University Press of America, 2012); Christine Kreamer et al., *African Cosmos: Stellar Arts* (National Museum of African Art, Smithsonian Institution; Monacelli Press, 2012).

43. David Owusu-Ansah, *Historical Dictionary of Ghana*, 4th ed., Historical Dictionaries of Africa (Rowman & Littlefield, 2014).

44. Martin Meredith, *The Fate of Africa: From the Hopes of Freedom to the Heart of Despair; a History of Fifty Years of Independence* (Public Affairs, 2005).

45. Crawford Young, *The Postcolonial State in Africa: Fifty Years of Independence, 1960–2010* (University of Wisconsin Press, 2012); Ivor Wilks, *Forests of Gold: Essays on the Akan and the Kingdom of Asante* (Ohio University Press, 1993); Jean Allman et al., *"I Will Not Eat Stone": A Women's History of Colonial Asante* (James Currey, 2000); R. S. Rattray, *The Tribes of the Ashanti Hinterland*, vol. 2, *Ashanti* (Clarendon Press, 1923).

46. Arjun Appadurai, *Modernity at Large: Cultural Dimensions of Globalization* (University of Minnesota Press, 1996).

47. Kwame Nkrumah, "Ghana Is Born," in *I Speak of Freedom* (Panaf, 1961), 107, as quoted in Nkrumah, "Independence Speech" (excerpt), in *The Ghana Reader: History, Culture, Politics*, ed. Kwasi Konadu and Clifford C. Campbell (Duke University Press, 2016), 301–2.

48. African Union, *Report of the Meeting of Experts from Member States on the Definition of the African Diaspora*, November 11, 2005, on.org/organs/ecossoc/Report-Expert-Diaspora_Defn_13april2005-Clean_copy1.doc; B. Holsey, *Routes of Remembrance: Refashioning the Slave Trade in Ghana* (University of Chicago Press, 2008); Jemima Pierre, *The Predicament of Blackness: Postcolonial Ghana and the Politics of Race* (University of Chicago Press, 2013).

49. World Bank, "Ghana Looks to Retool Its Economy as It Reaches Middle-Income Status," July 2011, https://www.worldbank.org/en/news/feature/2011/07/18/ghana-looks-to-retool-its-economy-as-it-reaches-middle-income-status.

50. Paul Gilroy, *The Black Atlantic: Modernity and Double Consciousness* (Harvard University Press, 1993); Holsey, *Routes of Remembrance*.

51. Matthew H. Brown, *Indirect Subjects: Nollywood's Local Address* (Duke University Press, 2021).

52. George Padmore, *Pan-Africanism or Communism: The Coming Struggle for Africa* (1956; Doubleday, 1971); Stephan Miescher and Leslie Ashbaugh, "Been-To Visions: Transnational Linkages Among a Ghanaian Dispersed Community in the Twentieth Century," *Ghana Studies Journal* 2 (1999): 5–95.

53. See Ghana Tourism homepage, accessed September 24, 2022, http://www.touringghana.com/mot.asp.

54. US Bureau of the Census, "American Community Survey 1-Year Data (2005–2022)," accessed September 1, 2024, https://www.census.gov/data/developers/data-sets/acs-1year.html; Takyiwaa Manuh, ed., *At Home in the World? International Migration and Development in Contemporary Ghana and West Africa* (Sub-Saharan Publishers, 2005); Ian E. A Yeboah, *Black African Neo-Diaspora: Ghanaian Immigrant Experiences in the Greater Cincinnati, Ohio Area* (Lexington Books, 2008).

55. Nettrice Gaskins, "Techno-Vernacular Creativity and Innovation across the African Diaspora and Global South," in *Captivating Technology: Race, Carceral Technoscience, and Liberatory Imagination in Everyday Life*, ed. Ruha Benjamin (Duke University Press, 2019), 252–74; Wiebe E. Bijker et al. eds., *The Social Construction of Technological Systems* (MIT Press, 1987).

56. Eric von Hippel, *Democratizing Innovation* (MIT Press, 2005).

57. Bruno Latour and Steve Woolgar, *Laboratory Life: The Construction of Scientific Facts* (Princeton University Press, 1986).

58. John Law, ed. *A Sociology of Monsters: Essays on Power, Technology, and Domination*, (New York: Routledge, 1991), 58–97; Leo Marx, "Technology: The Emergence of a Hazardous Concept," in *Technology and the Rest of Culture*, ed. Arien Mack (Ohio State University Press, 1997); Kevin Kelly, "The Computational Metaphor," in *The New Media Theory Reader*, ed. Robert Hassan and Julian Thomas (Open University Press, 1998).

59. In recent years, *applied science* has increasingly been used to name the interdisciplinary biological, physical, and engineering programs, such as Harvard's School of Engineering and Applied Sciences. The other term in current vogue is *STEM*, an acronym for science, technology, engineering, and mathematics (or medicine).

60. Castells, *Rise of Network Society*; Fred Turner, *From Counterculture to Cyberculture: Stewart Brand, the Whole Earth Network, and the Rise of Digital Utopianism* (University of Chicago Press, 2006).

61. Thomas P. Hughes, *Human-Built World: How to Think About Technology and Culture* (University of Chicago Press, 2004); Rosalind Williams, "Afterword: An Historian's View on the Network Society," in *The Network Society: A Cross-Cultural Perspective*, ed. Manuel Castells (Edward Elgar, 2004), 432–48.

62. Donald A. MacKenzie and Judy Wajcman, *The Social Shaping of Technology* (Open University Press, 1985); Bijker et al., *Social Construction of Technological Systems*; Bruno Latour, *Reassembling the Social: An Introduction to Actor-Network Theory* (Oxford University Press, 2005).

63. Jacques Ellul, *The Technological Society* (Vintage Books, 1954).

64. Max Weber, *Economy and Society: An Outline of Interpretive Sociology* (University of California Press, 1978).

65. Michael Adas, *Machines as the Measure of Men: Science, Technology, and Ideologies of Western Dominance* (Cornell University Press, 1990).

66. Marx, "Technology"; Williams, "Afterword."

67. Bijker et al., *Social Construction of Technological Systems*; MacKenzie and Wajcman, *Social Shaping of Technology*; Everett M. Rogers, *Diffusion of Innovations* (Free Press of Glencoe, 1962).

68. Ernest J. Wilson and Kevin R. Wong, eds., *Negotiating the Net in Africa: The Politics of Internet Diffusion* (Lynne Rienner, 2007).

69. Latour, "Technology Is Society Made Durable."

70. Payal Arora, *The Next Billion Users: Digital Life beyond the West* (Harvard University Press, 2019).

71. Larkin, *Signal and Noise*.

72. Rita Kiki Edozie, *"Pan" Africa Rising: The Cultural Political Economy of Nigeria's Afri-Capitalism and South Africa's Ubuntu Business*, Contemporary African Political Economy (Palgrave Macmillan, 2017).

73. Tejumola Olaniyan, "African Cultural Studies: Of Travels, Accents, and Epistemologies," in *Rethinking African Cultural Production*, ed. Frieda Ekotto and Kenneth W. Harrow (Indiana University Press, 2015); Keyan G. Tomaselli and Handel Kashope Wright, eds., *Africa, Cultural Studies and Difference* (Routledge, 2011).

74. Wilson and Wong, *Negotiating the Net in Africa*; Joshua Clark Davis, *From Head Shops to Whole Foods: The Rise and Fall of Activist Entrepreneurs*

(Columbia University Press, 2017).; Matthew H. Wisnioski, *Engineers for Change: Competing Visions of Technology in 1960s America*, Engineering Studies Series (MIT Press, 2012).

75. Justin Williams, *Pan-Africanism in Ghana: African Socialism, Neoliberalism, and Globalization* (Carolina Academic Press, 2016).

76. Andre Brock, "Critical Technocultural Discourse Analysis," *New Media & Society* 20, no. 3 (2018): 1012–30.

77. "Sheila O.," WPX Power 92, accessed April 19, 2022, https://www.power92chicago.com/show/sheila-o/.

78. Victoria Bernal, *Nation as Network: Diaspora, Cyberspace, and Citizenship* (University of Chicago Press, 2014).

79. Anna Everett, *Digital Diaspora: A Race for Cyberspace*, SUNY Series, Cultural Studies in Cinema/Video (SUNY Press, 2009).

80. Reginold Royston and Krystal Strong, "Reterritorializing Twitter: African Moments 2010-2015," in *#identity: Hashtagging Race, Gender, Sexuality, and Nation*, ed. Abigail De Kosnik and Keith Feldman (University of Michigan Press, 2019).

81. Hippel, *Democratizing Innovation*.

82. Reginold A. Royston, "At Home, Online: Affective-Exchange in Ghanaian Internet Video," in *Migrating the Black Body: Visual Culture and Diaspora*, ed. L. Raiford and H. Raphael-Hernandez (University of Washington Press, 2017).

83. Pew Research Center, "Mobile Fact Sheet," November 13, 2024, https://www.pewresearch.org/internet/fact-sheet/mobile/.

84. Yochai Benkler, *The Wealth of Networks: How Social Production Transforms Markets and Freedom* (Yale University Press, 2006).

85. Lynette Kvasny, "The Role of the Habitus in Shaping Discourses about the Digital Divide," *Journal of Computer-Mediated Communication* 10, no. 2 (January 1, 2005), https://doi.org/10.1111/j.1083-6101.2005.tb00242.x; Laura Robinson, "A Taste for the Necessary," *Information, Communication & Society* 12 (2009): 488–507.

86. Gilroy, *Black Atlantic*; M. Jacqui Alexander, *Pedagogies of Crossing* (Duke University Press, 2005); Joseph E. Harris, ed., *Global Dimensions of the African Diaspora* (Howard University Press, 1982).

87. Isidore Okpewho and Nkiru Nzegwu, eds., *The New African Diaspora* (Indiana University Press, 2009); Marilyn Halter and Violet Showers Johnson, *African & American: West Africans in Post-Civil Rights America* (New York University Press, 2016); Anima Adjepong, *Afropolitan Projects: Redefining Blackness, Sexualities, and Culture from Houston to Accra* (University of North Carolina Press, 2021).

88. Michel Laguerre, *Diaspora, Politics, and Globalization* (Palgrave Macmillan, 2006); Laura Candidatu et al., "Digital Diasporas: Beyond the Buzzword; Towards a Relational Understanding of Mobility and Connectivity," in *The Handbook of Diasporas, Media and Culture*, ed. Jessica Retis and Roza Tsagarousianou (Wiley-Blackwell, 2019).

89. The age of techne is described as a postindustrial world system, in contrast to the predigital periodization epoch the age of reason, a product of the European Enlightenment movement. Tom Boellerstorff, *Coming of Age*

in Second Life: An Anthropologist Explores the Virtually Human (Princeton University Press, 2008)

90. Claude Lévi-Strauss, *The Savage Mind* (University of Chicago Press, 1966).

91. For more on this, see the work of Mavhunga cited in note 40, as well as, for example Gloria Thomas Emeagwali and Edward Shizha, eds., *African Indigenous Knowledge and the Sciences: Journeys into the Past and Present* (Brill, 2016).

CHAPTER I. ASYMMETRICAL NETWORKS

1. Turner, *From Counterculture to Cyberculture.*

2. Thomas Friedman, *The World Is Flat: A Brief History of the Twenty-First Century* (Farrar, Straus and Giroux, 2005).

3. Larry Irving et al., "Falling through the Net: A Survey of the 'Have Nots' in Rural and Urban America," National Telecommunications and Information Administration (NTIA), July 1995, https://www.ntia.gov/page/falling-through -net-survey-have-nots-rural-and-urban-america; Dijk, *Digital Divide.*

4. Andrew Blum, *Tubes: A Journey to the Center of the Internet* (Harper-Collins, 2014).

5. There are more dense but shorter sea cables and segments between North Africa, Europe, and the Red Sea, but few of these lines connect to the east and west African coastal-spanning cables.

6. Jayne Miller, "Bandwidth and Pricing Trends in Africa: Patrick Christian at AfPIF 2018," *Telegeography* (blog), November 27, 2018, https://blog .telegeography.com/bandwidth-and-pricing-trends-in-africa-patrick-christian -at-afpif-2018.

7. Telegeography and the ITU reported in 2025 that Africa's used international bandwidth is estimated to be between 77 and 180 Tbs, while the largest international corridor going between North America and Europe was between 550 and 870 Tbs in volume. Telegeography, "Submarine Cable Map 2025," accessed April 25, 2025, https://submarine-cable-map-2025.telegeography.com/; ITU Data Hub, accessed April 25, 2025, https://datahub.itu.int/; Paul Brodsky, "Tracking International Traffic by Region," *Telegeography* (blog), October 12, 2023, https://blog.telegeography.com/tracking-international-internet-traffic-by -region.

8. Raley, *Tactical Media.*

9. Turner, *From Counterculture to* Cyberculture.

10. United Nations, *State of Broadband Report 2021* (Geneva: ITU and UNESCO, 2022).

11. Koen Leurs and Kevin Smets, "Five Questions for Digital Migration Studies: Learning from Digital Connectivity and Forced Migration In(to) Europe," *Social Media + Society* 4, no. 1 (January 2018): 1–16.

12. Ato Quayson, *Calibrations: Reading for the Social* (University of Minnesota Press, 2003), xv.

13. Ron Eglash et al., *Appropriating Technology: Vernacular Science and Social Power* (University of Minnesota Press, 2004).

14. See Magic Jack, accessed March 1, 2015, https://web.archive.org/web/20150103034320/https://www.magicjack.com/index.html.

15. Mantse and friends, interview with author, August 3, 2009, Oakland, California.

16. Mirca Madianou and Daniel Miller, *Migration and New Media: Transnational Families and Polymedia* (Routledge, 2012).

17. SMS refers to short message service, also called *texting*, a native program for mobile devices.

18. In comparison to the West, the sense that the Internet in Africa is *not* "always on" was borne out in my observations elsewhere in Africa over the years, specifically Senegal and Morocco, where Internet penetration has historically been high.

19. Officially the National Communications Authority (NCA), Ghana's telecom regulatory administration puts the number of mobile subscribers at 77 percent of the total population. Sometimes this *penetration number* is listed as over 100 percent of the population by some industry trackers, to account for the practice of *multi-SIMing* in the country, whereby an individual has multiple cellular numbers for distinctions between personal and work phones, as well as deploying a tactical approach to mobile phone user fees, which can be incentivized by different companies on any day of the week.

20. International Telecommunications Union (ITU), "Measuring Digital Development: State of Digital Development and Trends," 2025, Reports in the Africa, Americas, Asia/Pacific and European Regions, https://www.itu.int/hub/?s=State+of+digital+development+and+trends.

21. International Telecommunications Union (ITU), "Measuring Digital Development Facts and Figures," 2021, https://www.itu.int/en/ITU-D/Statistics/Documents/facts/FactsFigures2021.pdf.

22. 2G, 3G, and 4G refer to "generations" of mobile broadband technology, not gigabytes as in computing.

23. ITU, "Digital Technologies to Achieve the UN SDGs," May 31, 2022, https://www.itu.int:443/en/mediacentre/backgrounders/Pages/icts-to-achieve-the-united-nations-sustainable-development-goals.aspx.

24. June Arunga and Billy Kahora, *The Cell Phone Revolution in Kenya* (International Policy Network and Instituto Bruno Leoni, 2007), https://web.archive.org/web/20101128071345/http://www.policynetwork.net/development/publication/cell-phone-revolution-kenya.

25. Arora, *Next Billion Users*; United Nations, *State of Broadband Report: People Centered Approaches for Universal Broadband* (ITU and UNESCO, 2021), 5–6.

26. Appadurai, *Modernity at Large*.

27. Telegeography, *State of the Network*.

28. Often in the general public and even among non-Africa scholars, Africa's domination by Europe is thought to have begun with first contact. This is not the case. In fact, European geographic colonization did not start in earnest until the 19th century. Trade posts were the dominant interaction until the 19th century, and not until the 20th century did the majority of Africa experience the

kinds of colonization by White military, merchant, and missionary settlement that is more popularly understood.

29. Meredith, *Fate of Africa*, 2005; Walter Rodney, *How Europe Underdeveloped Africa* (Howard university Press, 1981).

30. Larkin, *Signal and Noise*.

31. Kevin Lewis O'Neill, "Disenfranchised: Mapping Red Zones in Guatemala City," *Environment and Planning A: Economy and Space* 51, no. 3 (May 2019): 654–69, https://doi.org/10.1177/0308518X18800069.

32. See chapter 4.

33. Valenzuela-Levi Nicolás, "The Written and Unwritten Rules of Internet Exclusion: Inequality, Institutions and Network Disadvantage in the Global South," *Communication & Society* 24, no. 11 (August 18, 2021): 1568–85.

34. Facebook's 2Africa underwater fiber optic line was landed in 2023. Csquared, a former Google partnership initially started in Ghana, will attempt to provide "open-access fiber-network" in the region. Google also initiated work on the Equiano network across West Africa in 2023.

35. USENet and Fido email networks had been in use for several years using these communications lines.

36. Wilson and Wong, *Negotiating the Net in Africa*.

37. In 2010 the Internet service providers (ISPs) MainOne and GLO established two more international cable connections, giving Ghana more than 5Tbs international bandwidth, dramatically increasing the capacity of users in the country.

38. "Life in the Fast Lane with the Bill Gates of Ghana," *Forbes Africa* (blog), [February 1, 2012], https://www.forbesafrica.com/entrepreneurs/2012/02/01/life-fast-lane-bill-gates-ghana/.

39. G. Pascal Zachary, "Black Star: Ghana, Information Technology and Development in Africa," *First Monday* 9, no. 3 (March 2004). https://firstmonday.org/ojs/index.php/fm/article/download/1126/1046/9786.

40. GSMA (formerly Groupe Spécial Mobile Association) is an industry consortium, data clearinghouse, and lobby, originally formed to advocate for the GSM standard in mobile telephony. It now serves as an organization for ISPs and represents over 750 firms.

41. Telegeography, *State of the Network*; GSMA, "The Mobile Economy: Sub-Saharan Africa 2018," 2018, http://www.gsma.com/r/mobileeconomy/.

42. This island nation hosts more Internet exchange and transfer points than any other locale in the African region.

43. Alliance for Affordable Internet, "Advancing Meaningful Connectivity: Towards Active & Participatory Digital Societies." May 24, 2022, https://a4ai.org/research-database/; United Nations, *2021 Annual Report: Reducing Inequality* (UNCTAD, 2021), https://unctad.org/annual-report-2021; Alexander Onukwue, "The Mobile Money Industry Processed More than $1 Trillion in 2021," *Quartz*, March 31, 2022, https://qz.com/africa/2149015/the-mobile-money-industry-processed-more-than-1-trillion-in-2021.

44. See the introduction, along with Crystal Brockton, "The Digital Divide in Ghana and Who's Closing the Gap," *Survive and Thrive* (blog), August 14,

2018, https://surviveandthriveboston.com/index.php/the-digital-divide-in-ghana
-and-whos-closing-the-gap/.

45. Herman Chinery-Hesse, interview with author, October 10, 2013.

46. Ibid.

47. Zachary, "Black Star."

48. Miescher, *Dam for Africa*.

49. The International Energy Association lists Ghana's energy consumption as doubling between the years 2006 and 2022, rising from 6.6 terawatt-hours to 13.5 TWh in that time period.

50. Miescher, *Dam for Africa*, 346.

51. Daily Guide, "Accra Floods: More than 100 Feared Dead after Explosion," *Modern Ghana*, July 4, 2015, https://www.modernghana.com/news/621226/accra -floods-more-than-100-feared-dead-after-explosion.html.

52. Meltwater Institute, "Homepage," MEST, January 18, 2024, https://melt water.org.

53. To jail break a device means to bypass manufacturer restrictions on device use through hacking or getting access to its "root" or primary coding architecture.

54. Onukwue, "Mobile Money." See more in chapter 5.

55. Madianou and Miller, *Migration and New Media*, 125.

56. Ushahidi has offices in Ghana and Kenya. The social impact platform was a pioneer in the use of crowdsourced data during crisis interventions in the early 2000s. See more at https://www.ushahidi.com.

57. Emmanuel, interview with author, October 29, 2013.

58. Gilles Deleuze and Félix Guattari, *A Thousand Plateaus: Capitalism and Schizophrenia* (University of Minnesota Press, 1987).

59. Castells, *Rise of the Network Society*, 501.

60. Pierre Lévy, *Cyberculture* (University of Minnesota Press, 2001), 113.

61. Castells actively disavowed public discourse advancing the "democratization" or "flattening" narrative of the Internet's capacities in the years after his influential writing in the 1990s, most notably in *Networks of Outrage and Hope: Social Movements in the Internet Age*, 2nd ed. (Polity Press, 2015).

62. Paul Baran, "On Distributed Communications Networks," *IEEE Transactions on Communications* 12, no. 1 (1964): 1–9.

63. ARPANET is the acronym for the US Department of Defense's Advanced Research Project Agency Network. This was one of the first transnational digital computer networks and an early "backbone" of the Internet.

64. Barry Leiner et al., "A Brief History of the Internet," The Internet Society, 1997, https://www.internetsociety.org/internet/history-internet/brief-history -internet/.

65. Castells, *Rise of the Network Society*, 501.

66. There is considerable debate about whether 25Mb/s download speed (aka 3G) should still be considered "high-speed broadband."

67. At the time I began this research in 2009, there were six national operators; by 2016 there were three additional urban broadband providers. As of 2022, five national carriers remained. Expresso had gone defunct in 2017, the

sole CDMA operator. Tigo (formerly Mobitel, Ghana's first cell phone company) and AirTel had merged. AirtelTigo, now the second-largest provider, was being acquired by the government in 2022 after going into financial arrears. Telecel Ghana was known as Vodafone Ghana until 2023, and I use this name in the context of my fieldwork.

68. R. Les Cottrell, "Pinging Africa: A Decadelong Quest Aims to Pinpoint the Internet Bottlenecks Holding Africa Back," *IEEE Spectrum* 50, no. 2 (2013): 54–59; GSMA, *State of Mobile Internet Connectivity 2019*; Alliance for Affordable Internet, "Advancing Meaningful Connectivity"; see World Bank Group, "Digital Economy for Africa Initiative," accessed April 20, 2024, https://www .worldbank.org/en/programs/all-africa-digital-transformation.

69. Bryan Pfaffenberger, "Social Anthropology of Technology," *Annual Review of Anthropology* 21 (1992): 491–516.

70. Pfaffenberger,, 499.

71. Ernesto Falcon, "Where Net Neutrality Is Today and What Comes Next," Electronic Frontier Foundation, December 28, 2021, https://www.eff.org /deeplinks/2021/12/where-net-neutrality-today-and-what-comes-next-2021 -review.

72. Terence K. Hopkins and Immanuel Maurice Wallerstein, *World-Systems Analysis: Theory and Methodology*, Explorations in the World-Economy (Sage Publications, 1982).

73. United Nations, *The Digital Economy Report 2024: Shaping an Environmentally Sustainable and Inclusive Digital Future* (United Nations, 2024), https://unctad.org/system/files/official-document/der2024_en.pdf.

74. Samir Amin, *Maldevelopment: Anatomy of a Global Failure* (Pambazuka, 2011); Rodney, *How Europe Underdeveloped Africa*.

75. James H. Smith, "Tantalus in the Digital Age: Coltan Ore, Temporal Dispossession, and 'Movement' in the Eastern Democratic Republic of the Congo," *American Ethnologist* 38, no. 1 (2011): 17–35.

CHAPTER 2. HACKING DEVELOPMENT

1. Mary Poppendieck and Tom Poppendieck, *Lean Software Development: An Agile Toolkit* (Addison-Wesley, 2003).

2. Richard Rottenburg, *Far-Fetched Facts: A Parable of Development Aid* (MIT Press, 2009); Gilbert Rist, *The History of Development: From Western Origins to Global Faith*, 4th ed. (Zed Books, 2014); Samir Amin, *Unequal Development: An Essay on the Social Formations of Peripheral Capitalism* (Monthly Review Press, 1976); Corrie Decker and Elizabeth McMahon, *The Idea of Development in Africa* (Cambridge University Press, 2021).

3. Tim Jordan and Paul A. Taylor, *Hacktivism and Cyberwars: Rebels with a Cause?* (Routledge, 2004).

4. Steven Levy. *Hackers: Heroes of the Computer Revolution*. Doubleday, 1984.

5. See more at Hack Hackers, https://www.hackshackers.com/about/. Google was tremendously influential in Accra during my time in the field. It supported the Vim! Series, paid speakers to attend Open Data events, and hosted an

annual Google Entrepreneurship Week in Ghana. In 2011 and 2012, Google Ghana personnel canvassed the middle-class and upper-income neighborhoods in Osu and Labone, offering to help small business owners get email and sign up for the social network Google+. At the same time, the University of Ghana was adopting Gmail as their internal communications service and encouraging the use of their free online tools for use in the classroom. They had recently hired Ashesi professors to research the potential impact of semi–smart phone adoption among working-class laborers in Accra with a free phone project. In the commercial field, Google was promoting an e-commerce marketplace, GoogleTrader-Africa, envisioned as a local alternative to Amazon.com. I interviewed several participants at hackathons in Ghana that participated in the Google Ambassadors program, which flies hundreds of students to a weeklong training and motivational conference.

6. Lilly Irani, "Hackathons and the Making of Entrepreneurial Citizenship," *Science, Technology, & Human Values* 40, no. 5 (September 2015): 799–824, https://doi.org/10.1177/0162243915578486.

7. Krystal Strong, "Do African Lives Matter to Black Lives Matter? Youth Uprisings and the Borders of Solidarity," *Urban Education* 53, no. 2 (February 2018): 265–85, https://doi.org/10.1177/0042085917747097.

8. Lindsey Whitfield, *Economies After Colonialism: Ghana and the Struggle for Power* (Cambridge University Press, 2020).

9. Jordan and Taylor, *Hacktivism and Cyberwars*.

10. McKenzie Wark, *A Hacker Manifesto* (Harvard University Press, 2022), 1.

11. Zachary, "Black Star."

12. Zeynep Tufekci, *Twitter and Tear Gas: The Power and Fragility of Networked Protest* (Yale University Press, 2017); Zayani, *Networked Publics and Digital Contention*.

13. *Hacker* is a colloquial term for a skilled computer programmer. Initially used in elite tech institutions and computer programmer circles as a superlative description for the best software programmers, the term took on an air of notoriety in the 1980s and 1990s, when underground computer techs began describing their illegal activities as hacking. In the current era, the terms *hacking* and *hackers* have been given a positive connotation by an emerging class of IT workers in Silicon Valley and throughout the world, describing highly skilled, efficient, and resourceful programming abilities and programmers. See Levy, *Hackers*.

14. Levy, *Hackers*.

15. DEFCON information is found at https://defcon.org/ (accessed September 1, 2018).

16. Douglas Thomas, *Hacker Culture* (University of Minnesota Press, 2002); Sarah R. Davies, "Characterizing Hacking: Mundane Engagement in US Hacker and Makerspaces," *Science, Technology, & Human Values* 43, no. 2 (March 2018): 171–97, https://doi.org/10.1177/0162243917703464.

17. My observations of hackathons are also based, in part, on my participation in hacker events in the San Francisco Bay Area as both an organizer and participant, as well as the research referenced. In the summer of 2011 I brought

a group of UC Berkeley students to Oakland, California's first civic-hack "Code4Oakland." In 2013 I was a co-organizer of University of California's first EduHack, attempting to produce tools for prospective college students.

18. MDGs were modified in 2015 to the more enduring sustainable development goals (SDGs). See United Nations, "Sustainable Development Goals," Department of Economic and Social Affairs Sustainable Development, accessed March 6, 2016, https://sdgs.un.org/goals.

19. Araba Sey, "'We Use It Different, Different': Making Sense of Trends in Mobil Phone Use in Ghana," *New Media & Society* 13, no. 3 (2011): 375–90.

20. Dorothea Kleine, *Technologies of Choice? ICTs, Development, and the Capabilities Approach* (MIT Press, 2013); Janet D. Kwami, "Development from the Margins? Mobile Technologies, Transnational Mobilities, and Livelihood Practices Among Ghanaian Women Traders," *Communication, Culture & Critique* 9, no. 1 (2016): 148–68, https://doi.org/10.1111/cccr.12136.

21. Jonathan Donner, "The Rules of Beeping: Exchanging Messages Via Intentional 'Missed Calls' on Mobile Phones," *Journal of Computer-Mediated Communication* 13, no. 1 (2007): 1–22, https://doi.org/10.1111/j.1083-6101 .2007.00383.x; see also Caribou Digital, "Jonathan Donner: Chief Knowledge Officer," accessed March 25, 2025, https://www.cariboudigital.net/advisor /jonathan-donner/.

22. Heather A. Horst and Daniel Miller, *The Cell Phone: An Anthropology of Communication* (Oxford & Berg Publishers, 2006); Arunga and Kahora, *Cell Phone Revolution in Kenya*.

23. Sey, "'We Use It Different, Different.'"

24. This is based on my monitoring of World Bank projects in Ghana between 2011 and 2018; data found at World Bank, "Archives: Ghana," accessed April 25, 2025, https://countryhistoricalprofiles.worldbank.org/?country=GHA&search profile=true.

25. Jonnie Akakpo and Mary H. Fontaine, "Ghana's Community Learning Centers," in *Telecentres: Case Studies and Key Issues; Management, Operations, Applications, Evaluation*, ed. C. Latchem and D. Walker (Commonwealth of Learning, 2001); Morten Falch and Amos Anyimadu, "Tele-Centres as a Way of Achieving Universal Access—the Case of Ghana," *Telecommunications Policy* 27, nos. 1–2 (February 2003): 21–39, https://doi.org/10.1016/S0308-5961(02) 00092-7; Ragnhild Overå, "Networks, Distance, and Trust: Telecommunications Development and Changing Trading Practices in Ghana," *World Development* 34, no. 7 (July 2006): 1301–15, https://doi.org/10.1016/j.worlddev.2005.11.015.

26. Don Slater and Janet D. Kwami, "Embeddedness and Escape: Internet and Mobile Use as Poverty Reduction Strategies in Ghana," 2005, Information Society Research Group Working Paper 4, https://www.researchgate.net /publication/228635823.

27. Wilson and Wong, *Negotiating the Net in Africa*.

28. *Redlining* historically refers to the practice of racial discrimination in the United States by real estate brokers and banks, seeking to keep African Americans out of White neighborhoods and business districts. For research on 419 scamming, called *sakawa* in Ghana, see Misty Bastian, "Nationalism in a Virtual Space: Immigrant Nigerians on the Internet," *West Africa Review* 23, no. 1, (1999): 1–9;

Jenna Burrell, *Invisible Users: Youth in the Internet Cafés of Urban Ghana* (MIT Press, 2012); James Yékú, *Cultural Netizenship: Social Media, Popular Culture, and Performance in Nigeria* (Indiana University Press, 2022).

29. HubTel, "Hubtel Ranked Ghana's Fastest Growing Company for 2022" (blogpost), May 16, 2024, https://blog.hubtel.com/ghanas-fastest-growing -company-for-2022/.

30. Seyram Avle et al., "Additional Labors of 2019the Entrepreneurial Self," *Proceedings of the ACM on Human-Computer Interaction* 3, no. 218 (November 7, 2019): 1–24, https://doi.org/10.1145/3359320.

31. "About Us," accessed January 2014, http://oasiswebsoft.com/aboutus.

32. AfriGadget is located at http://www.afrigadget.com. The site was created in 2006 by Kenyan-born Erik Hersman, who also blogs as "The White African," (whiteafrican.com), though he no longer contributes to AfriGadget. Hersman went on to be one of the founders of Kenya's internationally renowned tech incubator the iHub.

33. Pierre Lemonnier, *Elements for an Anthropology of Technology* (University of Michigan Press, 1992).

34. Jemila Abdulai, interview with author, January 13, 2015, via Google+ Hangouts.

35. STAR-Ghana (Strengthening Transparency Accountability and Responsiveness in Ghana). As an NGO, STAR-Ghana provided some training, framework agenda, and funds through USAID, the EU, the UK Department of International Development, and the Danish International Development Agency, with the goal "to increase the influence of civil society and Parliament in the governance of public goods and service delivery." See http://www.starghana.org, accessed January 8, 2013

36. Jemila Abdulai, "Ghana Decides Tag: Election 2012 Video Campaign," Ghana Decides, September 14, 2012, YouTube, https://www.youtube.com/watch ?v=3geg_CarGZU.

37. Wilson and Wong, *Negotiating the Net in Africa*.

38. Abdulai, Interview with author.

39. Ibid.

40. E. Gabriella Coleman, *Coding Freedom: The Ethics and Aesthetics of Hacking* (Princeton University Press, 2013), 14.

41. DIY refers to the phrase "do it yourself." Borrowing from the antiestablishment discourse of the US punk rock movement in the 1980s, DIY tinkerers and hackers espouse a culture of agency, or "maker culture," evident at events such as BarCamps and MakerFaires. See Amy Spencer, *DIY: The Rise of Lo-Fi Culture* (Marion Boyars, 2008); Kevin Wehr, *DIY: The Search for Control and Self-Reliance in the 21st Century* (Routledge, 2016).

42. BarCamp's California founders are no longer active in the community, but they maintain archival pages and a history of global BarCamps up to 2019. See "Organize a Local Bar Camp," accessed March 25, 2025, http://barcamp .org/w/page/404135/OrganizeALocalBarCamp.

43. GhanaThink Foundation, accessed November 10, 2013, and March 3, 2019, http://www.ghanathink.org/content/about-ghanathink. This website is no longer active, but the group maintains a web presence on LinkedIn.

44. Peter Geschiere et al, *Readings in Modernity in Africa* (Indian University Press, James Curry, Unisa Press, 2008).

45. successful SMS updates platform Esoko, run by BusyLabs, is named for the Swahili word *soko*, meaning "market," and thus not an indigenous word to Ghana.

46. George Jerry Swaniker, product developer for Subah Infosystems, interview with author, November 13, 2013, Accra. No relation to Fred Swaniker of the African Leadership Academy.

47. Kiesha Porter, "Dropifi Takes on Silicon Valley," *CNN*, July 10, 2013, http://www.cnn.com/2013/07/10/tech/web/ghana-dropifi-silicon-valley/.

48. Josh M., "Ghana Decides Diaspora Hangout on Election 2012," Ghana Decides, December 8, 2012, YouTube, https://www.youtube.com/watch?v=mmV-ML_PydI.

49. "Duapa Challenge—'The Campus Duel,'" British Council, accessed January 31, 2018, https://www.britishcouncil.org.gh/programmes/society/social-enterprise/duapa-challenge-campus-duel. The application Crowd'Frica discussed at the MPower payments event advanced to the international competition for Duapa in 2016.

50. Kwami, "Development from the Margins?"; Wallace Chipidza and Dorothy Leidner, "A Review of the ICT-Enabled Development Literature: Towards a Power Parity Theory of ICT4D," *Journal of Strategic Information Systems* 28, no. 2 (June 2019): 145–74, https://doi.org/10.1016/j.jsis.2019.01.002; Laura Schelenz and Maria Pawelec, "Information and Communication Technologies for Development (ICT4D) Critique," *Information Technology for Development* 28, no. 1 (January 2, 2022): 165–88, https://doi.org/10.1080/02681102.2021.1937473; Bidit Dey and Faizan Ali, "A Critical Review of the ICT for Development Research," in *ICTs in Developing Countries: Research, Practices and Policy Implications*, ed. Bidit Dey et al. (Palgrave Macmillan UK, 2016), https://doi.org/10.1057/9781137469502_1.

51. Jemima Pierre, "The Racial Vernaculars of Development: A View from West Africa," *American Anthropologist* 122, no. 1 (March 2020): 86–98.

52. Dambisa Moyo, *Dead Aid: Why Aid Is Not Working and How There Is a Better Way for Africa* (Farrar, Straus and Giroux, 2009).

53. Akua Akyaa Nkrumah, interview with author, November 29, 2016.

54. Akua Akyaa Nkrumah, "My Journey for Social Change: Confusion, Courage and Lots of Rubbish," December 13, 2016, YouTube, https://www.youtube.com/watch?v=zAiA1db6FBk.

55. Ibid.

56. Philip Mirvis and Bradley Googins, "Catalyzing Social Entrepreneurship in Africa: Roles for Western Universities, NGOs and Corporations," *Africa Journal of Management* 4, no. 1 (January 2, 2018): 57–83, https://doi.org/10.1080/23322373.2018.1428020; Uwafiokun Idemudia and Kenneth Amaeshi, eds., *Africapitalism: Sustainable Business and Development in Africa* (Routledge, 2019). see more in chapter 5.

57. United Nations, "Sustainable Development Goals."

58. William Senyo and Victor K. Ofoegbu, interview with author, November 29, 2016.

59. Patrick Awuah "How to Educate Leaders? Liberal Arts," June 2007, TedTalk, https://www.ted.com/talks/patrick_awuah_how_to_educate_leaders _liberal_arts.

60. Moyo, *Dead Aid.*

61. Miescher, *Dam for Africa,* 2022; Abena Dove Osseo-Asare, *Atomic Junction: Nuclear Power in Africa after Independence* (Cambridge University Press, 2019).

CHAPTER 3. DIGITAL DIASPORA

1. Some of the names used in this chapter are pseudonyms, developed in consultation with my research participants.

2. DJ Mantse, interview with author, October 10, 2009.

3. ITU, "World Telecommunication/ICT Indicators Database," International Telecommunications Union, accessed May 2014, https://www.itu.int/en/ITU-D /Statistics/Pages/publications/wtid.aspx.

4. Alexander, *Pedagogies of Crossing*; Brent Hayes Edwards, *The Practice of Diaspora: Literature, Translation, and the Rise of Black Internationalism* (Harvard University Press, 2003); Paul Gilroy, *The Black Atlantic: Modernity and Double Consciousness* (Harvard University Press, 1993).

5. Okpewho and Nzegwu, *New African Diaspora*; Halter and Johnson, *African & African American.*

6. Harris, *Global Dimensions of the African Diaspora.*

7. Edward Wilmot Blyden, *Christianity, Islam and the Negro Race* (1887; Black Classic Press, 1994.

8. W. E. B. Du Bois, *The World and Africa: An Inquiry into the Part Which Africa Has Played in World History* (International Publishers, 1947); Lorenzo Dow Turner, *Africanisms in the Gullah Dialect* (University of Chicago Press, 1949); Melville J. Herskovits, *The Myth of the Negro Past* (Harper and Bros., 1941).

9. Gilroy, *Black Atlantic.*

10. Ibid.

11. Okpewho and Nzegwu, *New African Diaspora*; Tejumola Olaniyan and James Sweet, eds., *The African Diaspora and the Disciplines* (Indiana University Press, 2010); Paul T. Zeleza, "African Diasporas: Towards a Global History," *African Studies Review* 53, no.1 (2020): 1–19.

12. Edwards, *Practice of Diaspora* ; Alexander, *Pedagogies of Crossing.*

13. Samantha Pinto, *Difficult Diasporas: The Transnational Feminist Aesthetic of the Black Atlantic* (New York University Press, 2013); Yomaira C. Figueroa-Vasquez, *Decolonizing Diasporas: Radical Mappings of Afro-Atlantic Literatures* (Evanston: Northwestern University Press, 2021).

14. William Safran, "Diasporas in Modern Societies: Myths of the Homeland and Return," *Diaspora: A Journal of Transnational Studies* 1, no. 1 (1991): 83–99; Robin Cohen, *Global Diasporas: An Introduction* (University of Washington Press, 1997); Rogers Brubaker, "The 'Diaspora' Diaspora," *Ethnic and Racial Studies* 28, no. 1 (2005): 1–9.

15. Deepika Bahri, "The Digital Diaspora: South Asians in the New Pax Electronica," in *In Diaspora: Theories, Histories, Texts*, ed. Makarand R. Paranjape (Indialog Publications, 2001).

16. Bonnie Nardi, *My Life as a Night Elf Priest: An Anthropological Account of World of Warcraft* (University of Michigan Press, 2010); John L. Jackson, *Thin Description: Ethnography and the African Hebrew Israelites of Jerusalem* (Harvard University Press, 2013).

17. Candidatu et al. "Digital Diasporas," 34.

18. Harris, *Global Dimensions of the African Diaspora*.

19. Steven Vertovec and Robin Cohen, eds., *Migration, Diasporas, and Transnationalism* (Edward Elgar, 1999), 3–4.

20. Harris, *Global Dimensions of the African Diaspora*.

21. George Shepperson, "African Diaspora: Concept and Context," in Harris, *Global Dimensions of the African Diaspora*, 52.

22. Robert Farris Thompson, *Flash of the Spirit: African and Afro-American Art and Philosophy* (Random House, 1983).

23. Khachig Tölöyan, 'The Nation-State and its Others: In Lieu of a Preface," *Diaspora: A Journal of Transnational Studies* 1, no. 1(1991): 3–7.

24. African Union, *Report of Meeting of Experts*.

25. Holsey, *Routes of Remembrance*.

26. Ibid.

27. Kevin K. Gaines, *American Africans in Ghana: Black Expatriates and the Civil Rights Era* (University of North Carolina Press, 2006); Maya Angelou, *All God's Children Need Traveling Shoes* (Vintage Books, 1991); Holsey, *Routes of Remembrance* ; Jemima Pierre, "'The Beacon of Hope for the Black Race': State Race-Craft and Identity Formation in Modern Ghana," *Cultural Dynamics* 21, no. 1 (2009): 29–50.

28. GhanaDiaspora.com was a site maintained by the Diaspora Support Unit in the Ministry of Foreign Affairs & Regional Integration in Ghana, accessed March 14, 2014, http://www.ghanaiandiaspora.com/. The motto of this website was "Think Home."

29. Kamari Maxine Clarke and Deborah A. Thomas, eds., *Globalization and Race: Transformations in the Cultural Production of Blackness* (Duke University Press, 2006), 133.

30. Ibid.

31. Cohen, *Global Diasporas*.

32. Campt, "The Crowded Space of Diaspora: Intercultural Address and the Tensions of Diasporic Relation," *Radical History Review* 83 (Spring 2002): 94–113; Law, *Sociology of Monsters*, 1991; Bruno Latour, *Reassembling the Social: An Introduction to Actor-Network Theory* (Oxford University Press, 2005).

33. Quayson, *Calibrations Reading for the Social*.

34. Steve Woolgar, "Configuring the User: The Case of Usability Trials," in *A Sociology of Monsters: Essays on Power, Technology, and Domination*, ed. John Law (Routledge, 1991).

35. Manuh, *At Home in the World?*

36. Dana Diminescu and Benjamin Loveluck, "Trances of Dispersion," *Crossings: Journal of Migration & Culture* 5, no. 1 (2014): 23–39; Laguerre, *Diaspora, Politics, and Globalization*; Bernal, *Nation as Network*.

37. Howard Rheingold, *The Virtual Community: Homesteading on the Electronic Frontier* (Addison-Wesley, 1993).

38. Multi-user dungeons (MUDs) were early electronic bulletin boards for gamer-centric communities.

39. The Akan word *okyeame* refers to the traditional "linguist" or aide to a ruler or king. This person interprets and speaks on behalf of the sovereign. The website Okyeame.net uses that title colloquially for its social network; see https://okyeame.net/okyeame/. NaijaNet, https://www.thenetnaija.com. See Everett, *Digital Diaspora*; Robert Tynes, "Nation-building and the Diaspora on Leonenet: A Case of Sierra Leone in Cyberspace," *New Media & Society* 9, no. 3 (2007): 497–518. Jennifer M. Brinkerhoff, "Digital Diasporas and Conflict Prevention: The Case of Somalinet.Com," *Review of International Studies* 32, no. 1 (January 2006): 25–47, https://doi.org/10.1017/S0260210506006917. Bernal, *Nation as Network*.

40. Madianou and Miller, *Migration and New Media*.

41. Ralph Ellison, *Shadow and Act* (1964; Vintage international, 1995).

42. Royston, "At Home, Online"; Reginold A. Royston, "Soulcraft: Theorizing Black Techne in African and American Viral Dance," *Social Media + Society* 8, no. 2 (April 2022), https://doi.org/10.1177/20563051221107644.

43. Janice Cheddie, "From Slaveship to Mothership and Beyond: Thoughts on a Digital Diaspora," in *Desire by Design: Body, Territories, and New Technologies*, ed. Cutting Edge (I. B. Tauris, 1999).

44. Ibid.

45. Ibid.

46. Eshun, "Further Considerations on Afrofuturism."

47. Daniel Miller and Don Slater, *The Internet: An Ethnographic Approach* (Berg Publishers, 2000).

48. Madianou and Miller, *Migration and New Media*.

49. Everett, *Digital Diaspora*.

50. Turner, *From Counterculture to Cyberculture*, 2006; Lévy, *Cyberculture*.

51. Castells, *Rise of Network Society*.

52. Political economy: Tufekci, *Twitter and Tear Gas*; Veronica Barassi, *Activism on the Web: Everyday Struggles Against Digital Capitalism* (Routledge, 2017); Marwan M. Kraidy, *Communication and Power in the Global Era: Orders and Borders* (Routledge, 2013). Race, culture, and gender: S. Craig Watkins, *Young and the Digital: What the Migration to Social-Network Sites, Games, and Anytime, Anywhere Media Means for Our Future* (Beacon Press, 2009); Noble, *Algorithms of Oppression*. Role of technical infrastructure: Ashwin J. Mathew, "The Myth of the Decentralised Internet," *Internet Policy Review* 5, no. 3 (September 30, 2016), https://doi.org/10.14763/2016.3.425.

53. Michel S. Laguerre, *The Multisited Nation: Crossborder Organizations, Transfrontier Infrastructure, and Global Digital Public Sphere* (Palgrave Macmillan, 2016).

54. Bernal, *Nation as Network*, 2.

55. Diaspora Affairs Bureau, "Diaspora Affairs Bureau," accessed May 1, 2014, https://diasporaaffairs.gov.gh; Ghana Statistical Service, *Ghana 2021 Population and Housing Census: Population of Regions and Districts*," 2021, https://statsghana.gov.gh/gssmain/fileUpload/pressrelease/2021%20PHC%20 General%20Report%20Vol%203A_Population%20of%20Regions%20and %20Districts_181121.pdf.

56. US Bureau of the Census, "American Community Survey," 2021, https:// data.census.gov/table/ACSDT5Y2020.B05006. In 2015 the Migration Policy Institute estimated there were more than 50,000 Ghanaians in the United States. See MPI, "The Ghanaian Diaspora in the United States," Rockefeller Aspen Diaspora Program, 2015, https://www.migrationpolicy.org/sites/default /files/publications/RAD-Ghana.pdf.

57. P. Owusu-Ankomah, "Emigration from Ghana: A Motor or Brake for Development" (presented at the 39th Session of the Commission on Population and Development, April 4, 2006), http://www.un.org/esa/population/cpd /cpd2006/CPD2006_Owusu_Ankomah_Statement.pdf.

58. Kwasi Konadu and Clifford C. Campbell, *The Ghana Reader: History, Culture, Politics* (Duke University Press, 2016).

59. Kwame Essien, *Brazilian-African Diaspora in Ghana: The Tabom, Slavery, Dissonance of Memory, Identity, and Locating Home* (Michigan State University Press, 2016).

60. David Kimble, *A Political History of Ghana: The Rise of Gold Coast Nationalism, 1850–1928* (Clarendon Press, 1963); Kofi Buenor Hadjor, *Nkrumah and Ghana: The Dilemma of Post-Colonial Power* (Kegan Paul International, 1988).

61. Takyiwaa Manuh, ed., " 'Efie' or the Meanings of 'Home' Among Female and Male Ghanaian Migrants in Toronto, Canada and Returned Migrants to Ghana," in *New African Diasporas*, ed. Khalid Koser (Taylor & Francis, 2009); John Dramani Mahama, *My First Coup d'Etat: Memories from the Lost Decades of Africa* (A&C Black, 2012).

62. Kwame Nkrumah, *Ghana: The Autobiography of Kwame Nkrumah* (Thomas Nelson and Sons, 1957).

63. Ibid.

64. Kwame Nimako, *Economic Change and Political Conflict in Ghana, 1600–1990* (Amsterdam: Thesis Publishers, 1991).

65. Ibid.; Jack Goody. *Technology, Tradition, and the State in Africa* (Oxford University Press, 1971).

66. MPI, Ghanaian Diaspora in the United States."

67. Manuh, *At Home in the World?*; Yeboah, *Black African Neo-Diaspora*; John A. Arthur, *The African Diaspora in the United States and Europe: The Ghanaian Experience* (Ashgate, 2008), http://public.eblib.com/EBLPublic/Public View.do?ptiID=438532.

68. Adams B. Bodomo et al., *The Okyeame Yearbook: A Transcript of Electronic Mail Discussions by a Group of Ghanaian Residents Abroad*, 1993, https:// okyeame.net/okyeame/okyeameyearbook.pdf; John Schaefer, "Discussion Lists

and Public Policy on iGhana," in *Native on the Net: Indigenous and Diasporic Peoples in the Virtual Age*, ed. K. Landzelius (Routledge, 2006).

69. Turner, *From Counterculture to Cyberculture*.

70. Rao, *Mobile Africa Report 2011*; GSMA, "Sub-Saharan Africa Mobile Economy 2013," *Groupe Speciale Mobile* (blog), November 1, 2011, http://www.gsma.com/mobileeconomyssa.

71. Karim H. Karim, *The Media of Diaspora* (Routledge, 2003), 12.

72. Takyiwaa Manuh, ed., *At Home in the World? International Migration and Development in Contemporary Ghana and West Africa* (Sub-Saharan Publishers, 2005); Arthur, *African Diaspora*.

73. Yeboah, *Black African Neo-Diaspora*, 50.

74. DJ Mantse, interview with author, March 8, 2014.

75. Clarke and Thomas, *Globalization and Race*.

76. music is a Ghanaian global pop music form.

77. DJ Mantse, interview with author, October 10, 2009.

78. Wendy Willems, "Beyond Platform-Centrism and Digital Universalism: The Relational Affordances of Mobile Social Media Publics," *Information, Communication & Society* 24, no.12 (2021): 1677–93.

79. Ama, interview with author, Amsterdam, October 15, 2013.

80. Rogers, *Diffusion of Innovations*.

81. John Elmina, interview with author, Sunnyvale, California, July 24, 2010.

82. GSMA, *The State of Mobile Internet Connectivity 2024*, October 2024, https://www.gsma.com/r/wp-content/uploads/2024/10/The-State-of-Mobile-Internet-Connectivity-Report-2024.pdf.

83. T-Mobile merged with Sprint to become the second-largest mobile carrier in the United States in 2020.

84. ITU, *Measuring the Information Society 2013*, 2013, https://www.itu.int/en/ITU-D/Statistics/Documents/publications/mis2013/MIS2013_without_Annex_4.pdf; GSMA, *The State of Mobile Internet Connectivity 2021*, 2021, https://www.gsma.com/r/wp-content/uploads/2021/09/The-State-of-Mobile-Internet-Connectivity-Report-2021.pdf; Alliance for Affordable Internet, "Advancing Meaningful Connectivity."

85. Sionne, interview with author, Chicago, July 10, 2016.

86. Ama, interview with author, October 15, 2013

87. Andoni Alonso and Pedro J. Oiarzabal, *Diasporas in the New Media Age: Identity, Politics, and Community* (University of Nevada Press, 2010); Leopoldina Fortunati et al., *Migration, Diaspora and Information Technology in Global Societies*, ed. Leopoldina Fortunati et al. (Routledge, 2012).

88. Larry Irving et al., "Falling through the Net: A Survey of the 'Have Nots' in Rural and Urban America," National Telecommunications and Information Administration (NTIA), July 1995, https://www.ntia.gov/page/falling-through-net-survey-have-nots-rural-and-urban-america; Watkins, *Young and the Digital*.

89. Royston and Strong, "Reterritorializing Twitter."

90. Media Ownership Monitor Ghana, "Bellaart Investments B.V. (GhanaWeb)," accessed July 3, 2021, https://ghana.mom-gmr.org/en/about/faq/.

91. Roberto Bezzicheri, interview with author, October 14, 2013. Bezzicheri left the company in 2021.

92. Kwadjo Yeboah, interview with author, October 16, 2013.

93. Robert Bellaart, interview with author, October 16, 2013.

94. As of 2024, it appears GhanaWeb is owned by Beta.Ventures, a Nigeria-based media holding company according to Media Ownership Monitor, 2021.

95. Lev Manovich, *The Language of New Media* (MIT Press, 2002).

96. Wendy Hui Kyong Chun, "The Enduring Ephemeral, or the Future Is a Memory," *Critical Inquiry* 35, no. 1 (2008): 148–71.

97. Michel S. Laguerre, *Urban Multiculturalism and Globalization in New York City: An Analysis of Diasporic Temporalities* (Palgrave Macmillan, 2003).

98. Edwards, *Practice of Diaspora*, 11–13.

99. Appadurai, *Modernity at Large*.

100. Victoria Bernal, "Nationalist Networks: The Eritriean Diaspora Online," in *Diasporas in the New Media Age*, ed. A. Alonso and P. J. Oiarzabal (University of Nevada Press, 2010).

101. Royston and Strong, "Reterritorializing Twitter."

102. Anima Adjepong, "Afropolitan Projects: African Immigrant Identities and Solidarities in the United States," *Ethnic and Racial Studies* 41, no. 2 (January 2018): 248–66, https://doi.org/10.1080/01419870.2017.1281985.

103. Lawrence Lessig, *The Future of Ideas: The Fate of the Commons in a Connected World* (Random House, 2001); Wendy Hui Kyong Chun and Thomas Keenan, *New Media, Old Media: A History and Theory Reader* (Psychology Press, 2006).

104. Kofi Aidoo, interview with author, Oakland, CA, July 20, 2009.

105. A. C. Eddie-Quartey, "The Story of GhanaFest-Chicago," Modern Ghana, August 12, 2005, https://www.modernghana.com/news/83919/the-story-of-ghanafest-chicago.html; Ghana National Council, in conversation with author, 2005.

106. Ghana National Council of Chicago, "Virtual Ghana Fest," Facebook, July 25, 2020, https://www.facebook.com/GNCCHICAGO/videos/344438430291924.

107. Omama TV in Chicago posts, via Facebook, https://www.facebook.com/kaakyiremarfo.

108. GNCC, "Virtual GhanaFest 2020," July 25, 2020, YouTube, https://www.youtube.com/watch?v=nNi9P9uToqg.

109. Nana "Nacee" Osei, in "Virtual GhanaFest 2020," YouTube, 3 hr., 57 min. to 4 hr., 6 min.

110. Royston, "At Home, Online."

CHAPTER 4. AFROPOLITAN MEDIASCAPES

1. Clifford Owusu (Kappacinco1), "US Visa Interview Gone Wrong," Two Lane Media, June 4, 2014, YouTube, https://youtu.be/lUqt5_jB8kk?si=qY3sbVSNKWxjPapm.

2. Jesse Weaver Shipley, "Transnational Circulation and Digital Fatigue in Ghana's Azonto Dance Craze," *American Ethnologist* 40 (2013): 362–81; see also Royston, "Soulcraft."

3. Kappacino1, "These new age kids songs hit different," TikTok, December 23, 2023, https://www.tiktok.com/@kappacinco1/video/7315825033937440042.

4. Clifford Owusu, interview with author via Skype, July 12, 2017.

5. Selasi Taiye, "Bye-Bye Babar: Or What Is an Afropolitan," *The LIP Magazine* (blog), March 3, 2005, https://thelip.robertsharp.co.uk/2005/03/03/bye-bye -barbar/.

6. Wisdom Tettey "Transnationalism, the African Diaspora, and the Deterritorialized Politics of the Internet," in *African Media and the Digital Public Sphere*, ed. Okoth Fred Mudhai et al. (Palgrave Macmillan, 2009), 144.

7. Appadurai, *Modernity at Large*, 8–9.

8. Castells, *Rise of the Network Society*, 408.

9. Stuart Hall, *Representation: Cultural Representations and Signifying Practices* (Sage and Open University, 1997).

10. Kamahra Ewing, "Transatlantic Connections: Reception and Production of Nollywood Films in Brazil," *Journal of African Cultural Studies* 31, no. 2 (May 4, 2019): 196–211.

11. Minna Salami, "Defining the Afropolitan," October 11, 2011, https://web .archive.org/web/20130513142910/http://www.ariselive.com/articles/defining -the-afropolitan/95788/.

12. "African, Dutch, or Afropolitan?," Radio Netherlands Worldwide (RNW), accessed June 13, 2013, http://www.rnw.nl/africa/article/african-dutch -or-afropolitan.

13. Mark Tutton, "Young, Urban and Culturally Savvy, Meet the Afropolitans," *CNN*, February 17, 2012, http://edition.cnn.com/2012/02/17/world /africa/who-are-afropolitans/index.html.

14. Stephanie Santana, "Exorcizing Afropolitanism: Binyavanga Wainaina Explains Why 'I Am a Pan-Africanist, Not an Afropolitan' at ASAUK 2012," Africa in Words, February 8, 2013, https://africainwords.com/2013/02/08 /exorcizing-afropolitanism-binyavanga-wainaina-explains-why-i-am-a-pan -africanist-not-an-afropolitan-at-asauk-2012/.

15. Adjepong, *Afropolitan Projects*, 16.

16. Ibid., 17.

17. Selasi, "Bye-Bye Babar."

18. Ibid.

19. Achille Mbembe, "Afropolitanism," in *Africa Remix: Contemporary Art of a Continent*, ed. Njami Simon and Lucy Durán (Jacana Media, 2007), 28–29.

20. Ibid., 28.

21. Ibid., 28.

22. Derrick Ashong, interview with author, February 18, 2013.

23. Santana, "Exorcizing Afropolitanism"

24. Rebecca Fasselt, "'I'm Not Afropolitan—I'm of the Continent': A Conversation with Yewande Omotoso," *The Journal of Commonwealth Literature* 50, no. 2 (June 2015): 231–46, https://doi.org/10.1177/0021989414552922.

25. Chimamanda N. Adichie, "Author Speaks," Books West Portal event, San Francisco, June 5, 2013.

26. Born in the US and variously raised between America and Africa, Davido has became a music celebrity synonymous with Afrobeats and Nigeria worldwide.

27. Franklin, Obeng-Odoom, "Africa: On the Rise, but to Where?," *Forum for Social Economics* 44, no. 3 (September 2, 2015): 234–50. https://doi.org/10 .1080/07360932.2014.955040.

28. The Afropolitan, http://www.afropolitan.co.za.

29. Joanna Wright, "Translating the 'Afropolitan,'" *The Media Online*, April 30, 2013, http://themediaonline.co.za/2013/04/translating-the-afropolitan/.

30. Gilroy, *Black Atlantic*.

31. Adi, *Pan-Africanism*; Imaobong Denis Umoren, *Race Women Internationalists: Activist-Intellectuals and Global Freedom Struggles* (University of California Press, 2018).

32. Ronald W. Walters, *Pan Africanism in the Africa Diaspora: An Analysis of Modern Afrocentric Political Movements* (Wayne State, 1997); Carmen M. White, "Living In Zion: Rastafarian Repatriates in Ghana, West Africa," *Journal of Black Studies* 37, no. 5 (May 2007): 677–709, https://doi.org/10.1177 /0021934705282379.

33. Manthia Diawara and Silvia Kolbowski, "Homeboy Cosmopolitan," *October* 83 (Winter 1998): 51–70; Mark Anthony Neal, *Looking for Leroy: Illegible Black Masculinities* (New York University Press, 2013).

34. Jemima Pierre, "Beyond Heritage Tourism," *Social Text* 27, no. 1 (2009): 59–81.

35. Adjepong, "Afropolitan Projects, ," 249.

36. Yogita Goyal, "When Was the Afropolitan? Thinking Literary Genealogy," *PMLA/Publications of the Modern Language Association of America* 136, no. 5 (October 2021): 782. https://doi.org/10.1632/S0030812921000675.

37. Taiye, "Bye-Bye Babar."

38. The Houston "Afropolitans" panel is archived at https://web.archive.org /web/20160303225012/www.hmaac.org/afropolitans.html, accessed November 15, 2013.

39. DNA and Soulfège, "Afropolitan," on *AFropolitan*, Ashong Ventures LLC, 2011.

40. Derrick Ashong, interview with author, February 18, 2013.

41. Simon Gikandi, "Foreword: On Afropolitanism," in *Negotiating Afropolitanism: Essays on Borders and Spaces in Contemporary African Literature and Folklore*, ed. Jennifer Wawrzinek and J. K. S. Makokha (Rodopi, 2011), 9–10.

42. Ahaspora Young Professionals, Ghana, Facebook, accessed October 9, 2013, https://www.facebook.com/ahaspora/; "About Ahaspora," accessed November 19, 2024, https://www.ahaspora.com/about-ahaspora/.

43. Amuzweni L. Ngoma, "South Africa's Black Middle Class Professionals" in *The Rise of Africa's Middle Class*, ed. Henning Melber (Zed Books, 2016).

44. Herman Chinery-Hess, interview with author, theSoftTribe headquarters, December 1, 2013.

45. Nicole Amarteifio, "An African City," August 5, 2014, YouTube, https://www.youtube.com/@AnAfricanCity/about.

46. The quote is taken from Nicole Amarteifio, "He Facebooked Me," An African City, April 4, 2014, YouTube, www.youtube.com/watch?v=1yKSTT-VhVM.

47. Yolanda Sangweni, "Ghanaian Writer Nicole Amarteifio on Creating the Hit Web Series, 'An African City,'" AfriPOP!, April 27, 2014, https://web.archive.org/web/20140414232158/http://afripopmag.com/2014/04/ghanaian-writer-nicole-amarteio-on-creating-the-hit-web-series-an-african-city/.

48. Appadurai, *Modernity at Large*, 8–9.

49. GhanaWeb can be accessed at https://www.ghanaweb.com/.

50. See chapter 3; see also "GhanaWeb at 22 Officially Launched in Accra," February 16, 2021, https://www.ghanaweb.com/GhanaHomePage/NewsArchive/GhanaWeb-at-22-officially-launched-in-Accra-1181554.

51. Mbembe, "Afropolitanism," 28–29.

52. Paul Carlucci, "Nigeria: Americanah," *Think Africa Press* (blog), May 17, 2013, https://allafrica.com/stories/201305200483.html.

53. Mwenya B. Kabwe, "Transgressing Boundaries: Making Theatre from an Afropolitan Perspective," *South African Theatre Journal* 21, no. 1 (January 2007): 48, https://doi.org/10.1080/10137548.2007.9687853.

54. Achille Mbembe, *On the Postcolony* (University of California Press, 2001); Ferguson, *Global Shadows*, 2006.

55. Ebenezer Forkuo Amankwaa et al., "'Away' Is a Place: The Impact of Electronic Waste Recycling on Blood Lead Levels in Ghana," *Science of the Total Environment* 601–2 (December 2017): 1566–74, https://doi.org/10.1016/j.scitotenv.2017.05.283.

56. Kamari Maxine Clarke, "Kony 2012, the ICC, and the Problem with the Peace-and-Justice Divide," *Proceedings of the ASIL Annual Meeting* 106 (2012): 309–13, https://doi.org/10.5305/procannmeetasil.106.0309.

57. Runako Celina et al., "Racism for Sale," *BBC: Africa Eye*, June 12, 2022, https://www.bbc.com/news/av/world-africa-61764466.

58. bell hooks, *Black Looks: Race and Representation* (South End Press, 1992).

59. Larkin, *Signal and Noise*.

60. Richard Dyer, *White* (Routledge, 1997).

61. Oluwakemi M. Balogun, "Beauty and the Bikini: Embodied Respectability in Nigerian Beauty Pageants," *African Studies Review* 62, no. 2 (June 2019): 80–102, https://doi.org/10.1017/asr.2018.125.

62. Burrell, *Invisible Users*, 189.

63. Katrien Pype, "Branhamist Kindoki: Ethnographic Notes on Connectivity, Technology, and Urban Witchcraft in Contemporary Kinshasa," in *Pentecostalism and Witchcraft*, ed. Knut Rio et al. (Springer, 2017), https://doi.org/10.1007/978-3-319-56068-7_5.

64. Yékú, *Cultural Netizenship*, 53.

65. Kwame Edwin Otu, "When the Lagoons Remember: An Afroqueer Futurist Reading of 'Blue Ecologies of Agitation,'" in "Gender and Sexuality in African Futurism," ed. Jacqueline-Bethel Tchouta Mougoué, special issue, *Feminist*

Africa 2, no. 2 (2021): 29–46; see also Reginold A. Royston, "From the Grammys to Agbogbloshie: African Viral Dance's Troubled Circuits," *The Black Scholar* 54, no. 3 (July 2, 2024): 11–18, https://doi.org/10.1080/00064246.2024.2364570.

66. Ben Asamoah et al., *Sakawa*, INTI Films, 2018, Vice News, The Short-list, February 15, 2021, YouTube, https://www.youtube.com/watch?v=ttT1hYR PD1k, 14 min., 42 sec. to 16 min.

67. Jennifer Wawrzinek, "Afropolitanism and the Novel: Mapping Material Networks in Recent Fiction from the African Diaspora," in *New Approaches to the Twenty-First-Century Anglophone Novel*, ed. Sibylle Baumbach and Birgit Neumann (Springer International Publishing, 2019), 253, https://doi.org/10.1007/978-3-030-32598-5_13.

68. Asamoah, *Sakawa*, 1 hr., 42 min., 44 sec. to 1 hr., 43 min., 30 sec.

69. Asamoah, *Sakawa*, 1 hr., 21 min., 01 sec.

70. ARRI, "Interview: Jonathan on Sakawa," February 2, 2020, YouTube, https://www.youtube.com/watch?v=L5JfCGS1FKY.

71. Stuart Hall, "Notes on Deconstructing 'the Popular,'" in *Foundations of Cultural Studies*, ed. David Morley et al. (Duke University Press, 2019), 355–56.

72. "Blitz the Ambassador's Afropolitan Dreams Block Party," *Afrofusion Lounge* (blog), May 7, 2013, http://afrofusionlounge.wordpress.com/2013/05/07/blitz-the-ambassadors-afropolitan-dreams-block-party/.

73. Blitz the Ambassador, "Africa Is the Future," on *Afropolitan Dreams* (Jakarta Records, 2014).

74. Goyal, "When Was the Afropolitan?."

75. Homi K. Bhabha, *The Location of Culture* (Routledge, 1994).

76. Yékú, *Cultural Netizenship*, 53

77. Halifu Osumare, *Hiplife in Ghana: West African Indigenization of Hip-Hop* (Palgrave Macmillan, 2013).

78. Melber, *Rise of Africa's Middle Class*.

79. Owusu, interview with author, July 12, 2017.

80. Ibid.

81. Gikandi, "foreward," 10.

82. The phrase was iconically used by Leonardo DiCaprio's Rhodesian mercenary character in *Blood Diamond* (2006); see Martha Evans and Ian Glenn, "'TIA—This Is Africa': Afropessimism in Twenty-First-Century Narrative Film," *Black Camera* 2, no. 1 (2010): 14, https://doi.org/10.2979/blc.2010.2.1.14.

83. A. M. Simone, *For the City Yet to Come: Changing African Life in Four Cities* (Duke University Press, 2004); Jordanna Matlon, *A Man Among Other Men: The Crisis of Black Masculinity in Racial Capitalism* (Cornell University Press, 2022).

84. Clark et al, *Pan African Spaces*, 10.

85. Castells, *Rise of the Network Society*, 408.

86. Chimamanda Ngozi Adichie, *Americanah* (Random House, 2013), 19.

87. Appadurai, *Modernity at Large*.

88. Kwame Anthony Appiah, *Cosmopolitanism: Ethics in a World of Strangers* (Penguin, 2015).

89. See chapter 3; see also Yeboah, *Black African Neo-Diaspora*.

90. Derrick Ashong, interview with author, San Francisco, California, February 18, 2013.

91. Daniela Nkechi Asante, interview with author, Amsterdam, NL via Zoom, August 26, 2022.

CHAPTER 5. PAN-AFRICANISM OR AFRICAPITALISM?

1. Some names have been changed in consultation with my research participants.

2. Arora, *Next Billion Users*.

3. Dayo Olopade, *The Bright Continent: Breaking Rules & Making Change in Modern Africa* (HMH, 2014); Moyo, *Dead Aid*; Pádraig Risteard Carmody, *The New Scramble for Africa* (Polity Press, 2011).

4. Moses E., Ochonu, ed., *Entrepreneurship in Africa: A Historical Approach* (Indiana University Press, 2018), 9; see also Eva Jordans et al., *Developing Global Leaders: Insights from African Case Studies* (Palgrave Macmillan, 2020), https://doi.org/10.1007/978-3-030-14606-1.

5. Karim Belayachi et al., "Entrepreneurship Database 2012," *Viewpoint: Public Policy for the Private Sector* 334 (October 2012). The database tracked entrepreneurship and development from 2006 to 2023.

6. Olayinka David-West et al., "Mobile Money as a Frugal Innovation for the Bottom of the Pyramid—Cases of Selected African Countries," *Africa Journal of Management* 5, no. 3 (2019): 274–302.

7. An abbreviation for mobile *pesa* or "money" in Swahili.

8. Reginold A. Royston, "Podcasts and New Orality in the African Mediascape," *New Media & Society* 25, no. 9 (September 2023): 2455–74, https://doi.org/10.1177/14614448211021032; see also Eric M. K. Osiakwan "The KINGS of Africa's Digital Economy," in *Digital Kenya: An Entrepreneurial Revolution in the Making*, ed. Bitange Ndemo and Tim Weiss (Springer, 2016).

9. World Bank, "Fintech for the Unbanked," Department of Economic and Social Affairs, Sustainable Development, accessed April 12, 2023, https://sdgs.un.org/partnerships/fintech-unbanked-promoting-financial-inclusion-poor-automobile-ownership-and-building.

10. GMSA, *The State of the Industry Report on Mobile Money 2024*, April 2024, https://www.gsma.com/sotir/wp-content/uploads/2024/03/GSMA-SOTIR-2024_Report_v7-2.pdf.

11. Mutsa Chironga et al., "Mobile Financial Services in Africa: Winning the Battle for the Customer," McKinsey & Company, September 1, 2017, https://wwww.mckinsey.com/industries/financial-services/our-insights/mobile-financial-services-in-africa-winning-the-battle-for-the-customer.

12. Girancourt et al., "How the COVID-19 Crisis May Affect Electronic Payments in Africa."

13. Georgina Maku Cobla and Eric Osei-Assibey, "Mobile Money Adoption and Spending Behaviour: The Case of Students in Ghana," *International Journal of Social Economics* 45, no. 1 (January 8, 2018): 29–42, https://doi.org/10.1108/IJSE-11-2016-0302.

14. Lily Kuo, "Africa Wasn't 'Rising' Before and It's Not 'Reeling' Now," Quartz Africa, October 2016, https://qz.com/africa/816006/africa-wasnt-rising -before-and-its-not-reeling-now; Ewout Frankema and Marlous van Waijenburg, "Africa Rising? A Historical Perspective," Africa Is a Country, October 17, 2018, https://africasacountry.com/2018/10/africa-rising-a-historical -perspective; Lorenzo Fioramonti, "The 'Africa Rising Story' Was Based on Faulty Logic—Here's How to Fix It," The Conversation, October 30, 2017, https://theconversation.com/the-africa-rising-story-was-based-on-faulty-logic -heres-how-to-fix-it-86327; Brahima S. Coulibaly, "In Defense of the 'Africa Rising' Narrative," Brookings Institute, June 27, 2017, https://www.brookings.edu /articles/in-defense-of-the-africa-rising-narrative/.

15. Pod Save Africa, Episode 41, "Death by Ambition," [2020], https:// soundcloud.com/user-682883785/reupload-episode-41-death-by-ambition; Freedom Jacob Caesar, "The African Dream—The Man with the Audacity to Build the New Africa," October 10, 2020, YouTube, https://youtu.be/CoI _y6cC9n4?si=SQ5l7B2_XRRyoHxl.

16. Jostein Hauge, "Africa's Economic 'Rise' Does Not Reflect Reality," *The Guardian*, September 3, 2014, https://www.theguardian.com/global -development/poverty-matters/2014/sep/03/africa-economic-rise-does-not -reflect-reality; Obeng-Odoom, "Africa."

17. Ato Ulzen-Appiah, interview with author, Accra, August 2023.

18. Victor Ofoegbu, interview with author, Accra, November 29, 2016.

19. For example, the MIT-Sloan African Innovate conference; Columbia University's long-running AfriCon; meetings such as the annual African Diaspora Network, based in Northern California; and Harvard University's 25-year-old African Business Conference.

20. For perspectives on the role of capitalist-focused development, see the work of Nana Osei-Opare; see also Thandika Mkandawire, "Thinking about Developmental States in Africa," *Cambridge Journal of Economics* 25, no. 3 (May 1, 2001): 289–314, https://doi.org/10.1093/cje/25.3.289.

21. Pierre, *Predicament of Blackness*.

22. Moyo, *Dead Aid*, xix.

23. In the vein of work from Walter Rodney and Immanuel Wallerstein.

24. Amin, *Unequal Development*; Rodney, *How Europe Underdeveloped Africa*; Mkandawire, "Thinking about Developmental States in Africa"; Wallerstein, "Africa in a Capitalist World"; Fernando Henrique Cardoso and Enzo Faletto, *Dependency and Development in Latin America* (University of California Press, 1979).

25. Pierre, "Racial Vernaculars of Development."

26. Nicholas Friederici et al., *Digital Entrepreneurship in Africa: How a Continent Is Escaping Silicon Valley's Long Shadow* (MIT Press, 2020); Akinyinka Akinyoade et al., eds., *Entrepreneurship in Africa* (Brill, 2017).

27. Established in 2019, the African Continental Free Trade Agreement (AfCFTA) is a multilateral treaty on tariffs and economic exchange binding the countries of the African Union and other regional economic communities. Its creation was considered a long-awaited triumph of Pan-Africanist financial

policy, as the treaty's aim is to create "common markets" among its member countries. More information is available at https://au-afcfta.org/about/. West Africa's ECOWAS is an economic and military association that has linked African states since 1975.

28. Richard Barbrook and Andy Cameron, "The Californian Ideology," *Science as Culture* 6, no. 1 (January 1996): 44–72, https://doi.org/10.1080/09505439609526455.

29. Olopade, *Bright Continent*.

30. See the annual reports from the Pan-African watchdog Afrobarometer, at https://www.afrobarometer.org/.

31. Ndemo and Weiss, *Digital Kenya*.

32. Edozie, *"Pan" Africa Rising*, 136.

33. Dotun Olowoporoku, interview with author via Skype, February 11, 2018.

34. Tony Elumelu and Heir Holdings, *The Africapitalist* 1, no. 1 (quarter 4, 2011): 1; Elumelu trademarked the concept in Nigeria but allows others to use it freely.

35. Tony Elumelu, "Africapitalism: The Path to Economic Prosperity and Social Wealth" (white paper, TEF, July 2022), https://www.tonyelumelufoundation.org/wp-content/uploads/dlm_uploads/2022/07/Africapitalism_The-Path-To-Economic-Prosperity-and-Social-Wealth.pdf. The firm's literature states that by 2007, United Bank for Africa "had grown from about 400 branches to more than 700, with more than 3,000 ATMs, up from 40 in 2005" (4).

36. Elumelu, *Africapitalist*, 1–2.

37. Adekunle Agbetiloye, "Tony Elumelu's UBA Surpasses ₦1 Trillion Market Capitalisation Mark," Business Insider, January 9, 2024, https://africa.businessinsider.com/local/markets/tony-elumelus-uba-surpasses-naira1-trillion-market-capitalisation-mark/b1d7s7v.

38. See the work of Mfuniselwa J. Bhengu, in Edozie, *"Pan" Africa Rising*.

39. The anti-debt Jubilee organization now operates under the name Debt Justice; see https://debtjustice.org.uk/.

40. Moyo, *Dead Aid*; K. Y. Amoako, *Know the Beginning Well: An Inside Journey Through Five Decades of African Development* (Africa World Press, 2020).

41. Elumelu, *Africapitalist*, 1.

42. Kenneth Amaeshi et al., eds., *Africapitalism: Rethinking the Role of Business in Africa* (Cambridge University Press, 2018).

43. Amartya Sen, *Development as Freedom* (Alfred Knopf, 1999).

44. For a more broadly considered exploration of CSR, see John Dunning, *Making Globalization Good: The Moral Challenges of Global Capitalism* (Oxford University Press, 2003).

45. Benedict Sheehy, "Defining CSR: Problems and Solutions," *Journal of Business Ethics* 131, no. 3 (October 2015): 625–48, https://doi.org/10.1007/s10551-014-2281-x.

46. Amaeshi et al., *Africapitalism*, 65.

47. Ibid.

48. Ibid., 71.

49. Stefan Ouma, "Africapitalism: A Critical Genealogy and Assessment," in, *Africapitalism: Sustainable Business and Development in Africa*, ed. Uwafiokin Idemudia and Kenneth Amaeshi (Routledge, 2019), 153.

50. Amaeshi et al., *Africapitalism*, 57.

51. C. W. Maris with Mogobe B. Ramose, "The Philosophy of Ubuntu and Ubuntu as Philosophy," in, *Philosophy from Africa: A Text with Readings,* 2nd ed., ed. Paul Coetzee and A. P. J. Roux (Oxford University Press, 2002).

52. José Cossa et al., "Cosmo-Ubuntu: Toward a New (Exterior to Modernity) Theorizing About the Human, the Cosmos, and Education," *Comparative Education Review* 64, no. 4 (2020): 753–56.

53. Rita Kiki Edozie, "Pan-Africanism Is Africa's Third Way: The Cultural Relevance of African Political Economy," in *The Palgrave Handbook of African Politics, Governance and Development*, ed. Samuel Ojo Oloruntoba and Toyin Falola (Palgrave Macmillan, 2018).

54. Elumelu, "Africapitalism," 6–8.

55. See chapter 2 and Pierre, "Racial Vernaculars of Development"; see also Diane Holt and David Littlewood, "Social Entrepreneurship and Africapitalism: Exploring the Connections," in *Africapitalism*, ed. Amaeshi et al. (Cambridge University Press, 2018), 235–66.

56. Paul Gifford, *Ghana's New Christianity, New Edition: Pentecostalism in a Globalising African Economy* (Indiana University Press, 2004).

57. Stephan Miescher, "The Akosombo Dam and the Quest for Rural Electrification in Ghana," in *Electric Worlds/Mondes Électriques: Creations, Circulations, Tensions, Transitions (19th–21st C.)*, Alain Beltran et al. (Peter Lang, 2016).

58. Osseo-Asare, *Atomic Junction*.

59. Mkandawire, "Thinking about Developmental States in Africa"; Meredith Woo-Cumings, ed., *The Developmental State* (Cornell University Press, 2019).

60. Nana Osei-Opare, "Ghana and Nkrumah Revisited: Lenin, State Capitalism, and Black Marxist Orbits," *Comparative Studies in Society and History* 65, no. 2 (April 2023): 416, https://doi.org/10.1017/S0010417522000548.

61. John Henrik Clarke and Amy Jacques Garvey, *Marcus Garvey and the Vision of Africa* (Random House, 1974).

62. Clarisse Burden-Stelly, *Black Scare, Red Scare: Theorizing Capitalist Racism in the United States* (University of Chicago Press, 2023).

63. Sören Scholvin and Moritz Breul, "An Unexpected Gateway: The Particularities of Mauritius as a Hub in Oil and Gas Gpns," *Development Southern Africa* 38, no. 1 (2021): 139–52; Will Fitzgibbon, "The Mauritius Leaks," International Consortium of Investigative Journalists, July 23, 2019, https://www.icij.org/investigations/mauritius-leaks/treasure-island-leak-reveals-how-mauritius-siphons-tax-from-poor-nations-to-benefit-elites/.

64. David Pilling, "How Covid-19 Interrupted the Mauritian Economic Miracle," *Financial Times,* April 1, 2022, https://www.ft.com/content/224ab32e-f84a-4eec-a57b-da73e841d010; see also Ramola Ramtohul and Thomas Hylland Eriksen, *The Mauritian Paradox: Fifty Years of Development, Diversity and Democracy* (Baltimore, MD: Project Muse/University of Mauritius Press, 2018).

65. Daniel Palmer, "Mauritius to License Crypto Custodians Starting in March," February 8, 2019, https://www.coindesk.com/markets/2019/02/08/mauritius-to -license-crypto-custodians-starting-in-march/.

66. Gagliardone, *China, Africa, and the Future of the Internet.*

67. Catherine A. Honeyman, *The Orderly Entrepreneur: Youth, Education, and Governance in Rwanda*, Anthropology of Policy (Stanford University Press, 2016).

68. Jess Auerbach Jahajeeah, "What a New University in Africa Is Doing to Decolonise Social Sciences," The Conversation, May 13, 2017, https://the conversation.com/what-a-new-university-in-africa-is-doing-to-decolonise-social -sciences-77181.

69. The Eurocentric focus of education in Africa is highlighted, for instance, in the recent controversies over the content and fairness of the West African Senior School Certificate Examination (WASSCE) and elsewhere in the British Commonwealth.

70. Pierre, "Racial Vernaculars of Development."; M. Neelika Jayaward- ane, "'The Capacity-Building-Workshop-in-Africa Hokum,'" *Journal of African Cultural Studies* 31, no. 3 (September 2, 2019): 276–80, https://doi.org/10.1080 /13696815.2019.1630265.

71. ALU's focus on "value-added" practical education is reminiscent of the historical distinction between area studies programs and Black studies pro- grams, marked by the "Black exodus" from the African Studies Association in the 1960–1970s led by John Henrik Clarke. See Jerry Gershenhorn, "'Not an Academic Affair': African American Scholars and the Development of African Studies Programs in the United States, 1942–1960," *Journal of African Ameri- can History* 94, no. 1 (2009): 44–68.

72. Freedom House, "Transnational Repression Is a Growing Threat to Global Democracy," June 4, 2021, https://freedomhouse.org/article/new-report -transnational-repression-growing-threat-global-democracy.

73. Reporters Without Borders, "Mauritius," accessed October 24, 2024, https://rsf.org/en/country/mauritius.

74. ICANN is a "multinational stakeholder group" that manages the assign- ment of website addresses and their numerical values. AFRINIC acts as a sub- sidiary group that does the same for African governments and polities.

75. Alan Suderman et al., "Africa Internet Riches Plundered, Contested by China Broker," *San Jose Mercury News*, September 30, 2021, https://www .mercurynews.com/2021/09/30/africa-internet-riches-plundered-contested-by -china-broker/.

76. ICANN, "Update on AFRINIC," September 22, 2023, https://www .icann.org/en/announcements/details/icann-update-on-afrinic-22-09-2023-en.

77. Suderman et al., "Africa Internet Riches Plundered."

78. In African public discourse, any nontitled individual under the age of 40 may be considered among the "youth," despite the continent's median age of 19. For more, see Krystal Strong and Christiana Kallon Kelly, "Youth Leadership for Development: Contradictions of Africa's Growing Leadership Pipeline," *The Journal of Modern African Studies* 60, no. 2 (June 2022): 217–38, https://doi .org/10.1017/S0022278X22000064.

79. Young African Leaders Initiative, "The Resilient Entrepreneur," US State Department, accessed August 1, 2024, https://yali.state.gov/courses/course-5170/#/lesson/lesson-1-planning-for-resilience.

80. African Union, "The African Smart Finance and Digital Banking Initiative," February 5, 2022, https://au.int/en/pressreleases/20220205/african-smart-finance-and-digital-banking-initiative-game-changer-msmes.

81. John C. Haltiwanger et al., "Who Creates Jobs? Small vs. Large vs. Young," *SSRN Electronic Journal*, August 1, 2010, US Census Bureau Center for Economic Studies Paper No. CES-WP- 10-17.

82. "Capitalism with a Conscience?," *African Business*, February 2, 2020, https://african.business/2020/02/economy/capitalism-with-a-conscience/.

83. Mkandawire, *African Intellectuals*, 6.

84. Royston, "Podcasts and New Orality in the African Mediascape."

85. George Ayittey, *Africa Unchained: The Blueprint for Africa's Future* (Palgrave Macmillan US, 2016).

86. Gracia Clark, *African Market Women: Seven Life Stories from Ghana* (Indiana University Press, 2010).

87. George Ayittey, "How Socialism Destroyed Africa," Heritage Foundation/Independent Institute, March 29, 2019, YouTube, https://www.youtube.com/watch?v=AEUUjoVjdMo.

88. Lee Wengraf, *Extracting Profit: Imperialism, Neoliberalism and the New Scramble for Africa* (Haymarket Books, 2018); see also Kingsley Ighobor, "Closing Africa's Wealth Gap," *Africa Renewal*, December 2017–March 2018, https://www.un.org/africarenewal/magazine/december-2017-march-2018/closing-africa%E2%80%99s-wealth-gap.

89. Samuel Ojo Oloruntoba, "Illicit Financial Flows and the African Development Conundrum," in *The Palgrave Handbook of African Politics, Governance and Development*, ed. Samuel Ojo Oloruntoba and Toyin Falola (Palgrave Macmillan, 2018).

90. For example, Naspers, a South Africa-Dutch conglomerate and former apartheid government collaborator, owns significant media programming content outlets in Ghana, including the popular movie channel AfricaMagic and the cable provider DSTV. The majority of underwater fiber optic sea cable companies are owned by non-Africans or have African conglomerates as minor partners. Asoko Insight, "Africa's Largest Telecommunications Companies" (blog post), August 18, 2022, https://www.asokoinsight.com/content/market-insights/africa-largest-telecoms-companies.

91. Ciku Kimeria, "A White Founder's $1 Million Nairobi Food Startup Aims to Fix a Problem Kenyans Say Doesn't Exist," Quartz, June 25, 2021, https://qz.com/africa/2024177/white-privilege-in-the-african-startup-scene-draws-ire-of-kenyans; Muthuri Kinyamu (@MuthuriKinyamu), "Silicon Savannah 2018 or should we call it business community?" Twitter/X, August 27, 2018, 10:33 a.m., https://x.com/MuthuriKinyamu/status/1034101487419510784?mx=2.

92. "Dangote and Elumelu Under Fire for Attempting to Corner End-SARS Protesters," *Business Elites Africa*, October 14, 2020, https://businesselitesafrica.com/dangote-elumelu-under-fire-for-attempting-to-corner-endsars-protesters/.

93. Moses E. Ochonu, "The Dangote Paradox," *Sahara Reporters* (blog), March 25, 2016, https://saharareporters.com/2016/03/25/dangote-paradox-moses-e-ochonu.

94. Ouma, "Africapitalism," 150.

95. Roger Southall, *The New Black Middle Class in South Africa* (James Currey, 2016).

96. Edgar Cooke et al., *The Ghana Poverty and Inequality Report: Using the 6th Ghana Living Standards Survey* (University of Sussex, 2016), 1–43; Opeoluwa Adeniyi Adeosun and Mosab I. Tabash, "Pro-Poor and Inclusive Growth in West Africa," *African Journal of Economic and Management Studies* 13, no. 1 (2022): 105–35.

97. Roland Amoah et al., *Building an Inclusive Digital Payments Ecosystem: The Way Forward* (United Nations and Better Than Cash Alliance, 2017), https://www.betterthancash.org/alliance-reports/country-diagnostic-ghana.

98. Sam Dudek with Kevin Bannerman-Hutchful, "Positive Effects of Mobile Money on Financial Inclusivity," *Equilibrium: Undergraduate Journal of Economics* 8 (June 8, 2018); Reginold A. Royston, "Exploring Mobile Money in Accra 2017" (unpublished ms., African Global Ties project, n.d.), https://www. raroyston.com.

99. We also interviewed mobile money agents, local top-up sellers, and accountants who could dispense cash to users from digital wallets.

100. Loraine, interview with research assistant Kevin Bannerman-Hutchful, January 28, 2017, East Legon.

101. Caroline, interview with research assistant Kevin Bannerman-Hutchful, January 29, 2017, Adabraka.

102. Robert Darko Osei et al., *Structural Transformation and Inclusive Growth in Ghana*, 37th ed. (United Nations University-WIDER, 2020), https://doi.org/10.35188/UNU-WIDER/2020/794-1.

103. Gaia Penteriani, *The e-Levy in Ghana: Economic Impact Assessment.* GSMA, March 2023. https://www.gsma.com/publicpolicy/wp-content/uploads/2023/03/E-Levy-Ghana-Economic-Impact-Assessment.pdf, 8.

104. Steven J. L. Taylor, *Exiles, Entrepreneurs, and Educators: African Americans in Ghana* (SUNY Press, 2019).

105. New Patriotic Party (NPP), "Manifesto 2020: Leadership of Service," accessed December 9, 2020, https://media.peacefmonline.com/docs/202008/353419888_327383.pdf.

106. Adela Suliman, "Africans Must Move Beyond Aid and Not Be 'Beggars of the World,'" Reuters, November 21, 2017, https://www.reuters.com/article/idUSKBN1DL2JB/.

107. Penteriani, *E-Levy in Ghana.*

108. Covid-19 Health Recovery Levy Act of 2021, March 31, 2021.

109. Smith Oduro-Marfo, "Transient Crisis, Permanent Registries" in *Data Justice and Covid-19: Global Perspectives*, ed. Linnet Taylor et al. (Meatspace Press, 2020), https://globaldatajustice.org/data-justice-and-covid-19/, 143.

110. Royston and Strong, "Reterritorializing Twitter."

111. Sena Alico with Joshua Boye-Doe, "#FixtheCountry—Youth in Politics," *The Progressive Minds Show*, Facebook, May 16, 2021, https://www.facebook.com/watch/live/?ref=watch_permalink&v=2932585950331373.

112. MyJoyOnline, "Annoh-Dompreh Apologises for #FixYourself Comment," May 4, 2021, https://www.myjoyonline.com/annoh-dompreh-apologises-for-fixyourself-comment/.

113. Agana-Nsiire Agana and Charles Prempeh, "Of Farms, Legends, and Fools: Re-Engaging Ghana's Development Narrative Through Social Media," *Media, Culture & Society* 44, no. 7 (October 2022): 1290–1306, https://doi.org/10.1177/01634437221111918.

114. Marian Ansah, "#FixtheCountry Campaigners to Protest Over Assault on Caleb Kudah and Other Journalists," CitiNews, May 17, 2001, https://citinewsroom.com/2021/05/fixthecountry-campaigners-to-protest-over-assault-on-caleb-kudah-and-other-journalists/.

115. Macho Kaaka, "We Will Not Vote for Any One," Facebook, June 24, 2021, https://www.facebook.com/macho.kaaka/posts/566396964766060.

116. Freedom House, "Ghana," 2022, https://freedomhouse.org/country/ghana/freedom-net/2022.

117. Mahmud Mohammed-Nurudeen, "Ejura Shooting: 2 Families Receive ₵500k Compensation from Government," MyJoyOnline, February 6, 2022, https://www.myjoyonline.com/ejura-shooting-2-families-receive-%c2%a2500k-compensation-from-government/.

118. The Gã phrase *julorbi* means "child of a thief." The hashtag satirizes the name of the Ghanaian president's residence and office, Jubilee House.

119. Emmanuel Akinwotu, "Ghana 'Fix the Country' Activist Says He Was Assaulted and Illegally Detained," *The Guardian*, July 14, 2022, https://www.theguardian.com/world/2022/jul/14/ghana-activist-oliver-barker-vormawor-lawsuit-arrest-detention.

120. Nana Oye Ankrah and Alexis Akwagyiram, "Who Is Ghana's Masked Presidential Contender?," *Semafor Africa*, December 8, 2023, https://www.semafor.com/article/12/08/2023/ghana-masked-presidential-contender.

121. The full address can be found at Joy News, "New Africa Foundation Convention Cancelled, Main Speakers Hold a News Conference," January 8, 2024, YouTube, https://www.youtube.com/watch?v=77-wx4lcHMQ.

122. Edozie, "Pa-Africanism Is Africa's Third Way," 114.

123. Jesse Weaver Shipley, "The Market Decides If We Are Free," Africa Is a Country, January 16, 2017, https://africasacountry.com/2017/01/the-market-decides-if-we-are-free.

124. As of 2024, #FixtheCountry organizer Kalyjay resides in Texas.

125. Padmore, *Pan-Africanism or Communism*, 379.

126. Jodi Dean, "Neofeudalism: The End of Capitalism?," *Los Angeles Review of Books*, May 12, 2020, https://lareviewofbooks.org/article/neofeudalism-the-end-of-capitalism/.

127. Amin, *Maldevelopment*.

128. Ruth Olurounbi, "Nigeria Central Bank Orders Cryptocurrency Accounts to Close," Bloomberg, February 5, 2021, https://www.bloomberg.com/news/articles/2021-02-05/nigerian-central-bank-orders-closure-of-cryptocurrency-accounts; see also David Golumbia, *The Politics of Bitcoin: Software as Right-Wing Extremism* (University of Minnesota Press, 2016).

CONCLUSION

1. ITU, *The ICT Development Index,* 2024 (United Nations, June 2024), https://www.itu.int/itu-d/reports/statistics/idi2024/.

2. Ibid.

3. Among others, see Zayani, *Networked Publics and Digital Contention*; Msia Kibona Clark and Wunpini Fatimata Mohammed, eds., *African Women in Digital Spaces: Redefining Social Movements on the Continent and in the Diaspora* (Mkuki na Nyota, 2023); Thiam and Rochon, *Sustainability, Emerging Technologies, and Pan-Africanism*; Mutsvairo, *Digital Activism in the Social Media Era*; Dunn et al, *Re-Imagining Communication in Africa and the Caribbean.*

4. Mavhunga, *What Do Science, Technology, and Innovation Mean?*, 1.

5. The Akan phrase *sankofa* refers to the proverb, "If you have forgotten something, it is okay to go back and retrieve it," often interpreted to mean that one cannot move forward without remembering one's past.

6. For more on their work, see Tawia's memoir: Eyram Tawia, *Uncompromising Passion: The Humble Beginnings of an African Video Game Industry* (BKC Consulting, 2016); Kwabena Opoku-Agyemang, "Lost/Gained in Translation: Oware 3D, Ananse; the Origin and Questions of Hegemony," *Journal of Gaming & Virtual Worlds* 7, no. 2 (June 1, 2015): 155–68, https://doi.org/10.1386/jgvw.7.2.155_1.

7. The Transatlantic Slave Trade started in the early 16th century, and the first enslaved Africans were transported to Europe more than 500 years ago.

8. The African Diaspora Development Institute, https://ouraddi.org/about-addi/.

9. The term refers to the confluence of artificial intelligence, robotic automation, and biotech's role in transforming work and the labor market in the 21st century; see Klaus Schwab, *The Fourth Industrial Revolution* (Crown Business, 2017).

10. Kofi B. Asante, "Nkrumah and State Enterprises," in *The Life and Work of Kwame Nkrumah: Papers of a Symposium Organized by the Institute of African Studies, University of Ghana, Legon,* ed. Kwame Arhin (Accra: Sedeo, 1991), 253.

11. new media discourse, *the singularity* refers to a moment when the confluence of technology and human agency surpasses conventional projects of historical continuity, ushering in a utopian era. This is usually imagined as the emergence of conscious artificial intelligence. See Murray Shanahan, *The Technological Singularity* (MIT Press, 2015).

12. Global Startup Ecosystem, "Ghana Tech Summit (Digital)," December 10, 2020, https://www.youtube.com/live/Q1eHOaCyDiI?si=Q80AJiF2F3oNpozo, 43 min., 11 sec.

13. Ibid., 1 hr., 1 min. to 1 hr., 32 min.

14. Non-fungible tokens (NFTs) are unique digital signatures (art pieces in this case) utilizing "blockchain" encryption tools.

15. CoinDesk, "Afropolitan Raises $2.1M to Build a Digital Nation," First Mover, September 13, 2022, YouTube, https://www.youtube.com/watch?v=vLjooX1irMI.

Bibliography

Blogs, social media posts, and home pages cited in the notes are not included here.

Adas, Michael. *Machines as the Measure of Men: Science, Technology, and Ideologies of Western Dominance*. Cornell University Press, 1990.

Adeosun, Opeoluwa Adeniyi, and Mosab I. Tabash. "Pro-Poor and Inclusive Growth in West Africa." *African Journal of Economic and Management Studies* 13, no. 1 (2022): 105–35.

Adi, Hakim. *Pan-Africanism: A History*. Bloomsbury Academic, 2018.

Adichie, Chimamanda Ngozi. *Americanah*. Random House, 2013.

Adjepong, Anima. "Afropolitan Projects: African Immigrant Identities and Solidarities in the United States." *Ethnic and Racial Studies* 41, no. 2 (January 2018): 248–66. https://doi.org/10.1080/01419870.2017.1281985.

———. *Afropolitan Projects: Redefining Blackness, Sexualities, and Culture from Houston to Accra*. University of North Carolina Press, 2021.

African Business. "Capitalism with a Conscience?" February 2, 2020. https://african.business/2020/02/economy/capitalism-with-a-conscience/.

African Diaspora Development Initiative. "Wakanda One: The City of Return." Accessed July 15, 2021. https://ouraddi.org/wakanda-one/

"African, Dutch, or Afropolitan?" Radio Netherlands Worldwide (RNW). Accessed June 13, 2013. http://www.rnw.nl/africa/article/african-dutch-or-afropolitan.

African Union. "The African Smart Finance and Digital Banking Initiative." February 5, 2022. https://au.int/en/pressreleases/20220205/african-smart-finance-and-digital-banking-initiative-game-changer-msmes.

———. *Report of the Meeting of Experts from Member States on the Definition of the African Diaspora*. November 11, 2005. on.org/organs/ecossoc/Report-Expert-Diaspora_Defn_13april2005-Clean_copy1.doc.

Agana, Agana-Nsiire, and Charles Prempeh. "Of Farms, Legends, and Fools: Re-Engaging Ghana's Development Narrative Through Social Media." *Media, Culture & Society* 44, no. 7 (October 2022): 1290–1306. https://doi.org/10.1177/01634437221111918.

Agbetiloye, Adekunle. "Tony Elumelu's UBA Surpasses ₦1 Trillion Market Capitalisation Mark." Business Insider, January 9, 2024. https://africa.business insider.com/local/markets/tony-elumelus-uba-surpasses-naira1-trillion -market-capitalisation-mark/b1d7s7v.

Akakpo, Jonnie, and Mary H. Fontaine. "Ghana's Community Learning Centers." In *Telecentres: Case Studies and Key Issues; Management, Operations, Applications, Evaluation*, edited by C. Latchem and D. Walker. Commonwealth of Learning, 2001.

Akinwotu, Emmanuel. "Ghana 'Fix the Country' Activist Says He Was Assaulted and Illegally Detained." *The Guardian*, July 14, 2022. https://www .theguardian.com/world/2022/jul/14/ghana-activist-oliver-barker-vormawor -lawsuit-arrest-detention.

Akinyoade, Akinyinka, Ton Dietz, and Chibuike Uche, eds. *Entrepreneurship in Africa*. African Dynamics. Brill, 2017.

Alexander, M. Jacqui. *Pedagogies of Crossing: Meditations on Feminism, Sexual Politics, Memory, and the Sacred*. Duke University Press, 2005.

Alliance for Affordable Internet. "Advancing Meaningful Connectivity: Towards Active & Participatory Digital Societies." May 24, 2022. https://a4ai.org /research-database/.

Allman, Jean Marie, Victoria Tashjian, and Victoria B. Tashjian. *"I Will Not Eat Stone": A Women's History of Colonial Asante*. Heinemann, 2000.

Alonso, Andoni, and Pedro J. Oiarzabal. *Diasporas in the New Media Age: Identity, Politics, and Community*. University of Nevada Press, 2010.

Amaeshi, Kenneth, Adun Okupe, and Uwafiokun Idemudia, eds. *Africapitalism: Rethinking the Role of Business in Africa*. Cambridge University Press, 2018.

Amankwaa, Ebenezer Forkuo, Kwame A. Adovor Tsikudo, and Jay A. Bowman. "'Away' Is a Place: The Impact of Electronic Waste Recycling on Blood Lead Levels in Ghana." *Science of the Total Environment* 601–2 (December 2017): 1566–74. https://doi.org/10.1016/j.scitotenv.2017.05.283.

Amin, Samir. *Maldevelopment: Anatomy of a Global Failure*. Pambazuka, 2011.

———. *Unequal Development: An Essay on the Social Formations of Peripheral Capitalism*. Monthly Review Press, 1976.

Amoah, Roland, Rajesh Bansal, Aneth Kasebele, and Ariadne Plaitakis. *Building an Inclusive Digital Payments Ecosystem: The Way Forward*. United Nations and Better Than Cash Alliance, 2017. https://www.betterthancash .org/alliance-reports/country-diagnostic-ghana.

Amoako, K. Y. *Know the Beginning Well: An inside Journey Through Five Decades of African Development*. Africa World Press, 2020.

Anderson, Reynaldo, and Charles E. Jones, eds. *Afrofuturism 2.0: The Rise of Astro-Blackness*. Lexington Books, 2016.

Angelou, Maya. *All God's Children Need Traveling Shoes*. Vintage Books, 1991.

Ankrah, Nana Oye, and Alexis Akwagyiram. "Who Is Ghana's Masked Presidential Contender?" *Semafor Africa*, December 8, 2023. https://www.semafor.com/article/12/08/2023/ghana-masked-presidential-contender.

Ansah, Marian. "#FixtheCountry Campaigners to Protest Over Assault on Caleb Kudah and Other Journalists." CitiNews. May 17, 2001. https://citinewsroom.com/2021/05/fixthecountry-campaigners-to-protest-over-assault-on-caleb-kudah-and-other-journalists/.

Appadurai, Arjun. *Modernity at Large: Cultural Dimensions of Globalization*. University of Minnesota Press, 1996.

Appiah, Ato Ulzen. Presentation at Ghana Tech Summit, Accra International Conference Centre, December 13–15, 2019.

Appiah, Kwame Anthony. *Cosmopolitanism: Ethics in a World of Strangers*. Penguin, 2015.

Arora, Payal. *The Next Billion Users: Digital Life Beyond the West*. Harvard University Press, 2019.

Arthur, John A. *The African Diaspora in the United States and Europe: The Ghanaian Experience*. Ashgate, 2008. http://public.eblib.com/EBLPublic/PublicView.do?ptiID=438532.

Arunga, June, and Billy Kahora. *The Cell Phone Revolution in Kenya*. International Policy Network and Instituto Bruno Leoni, 2007. https://web.archive.org/web/20101128071345/http://www.policynetwork.net/development/publication/cell-phone-revolution-kenya.

Asante, Kofi B. "Nkrumah and State Enterprises." In *The Life and Work of Kwame Nkrumah: Papers of a Symposium Organized by the Institute of African Studies, University of Ghana, Legon*, edited by Kwame Arhin. Accra: Sedeo, 1991.

Avle, Seyram. "Articulating and Enacting Development: Skilled Returnees in Ghana's ICT Industry." *Information Technologies and International Development* 10 (December 2014): 1–13.

Avle, Seyram, Julie Hui, Silvia Lindtner, and Tawanna Dillahunt. "Additional Labors of the Entrepreneurial Self." *Proceedings of the ACM on Human-Computer Interaction* 3, no. CSCW (November 7, 2019): 1–24. https://doi.org/10.1145/3359320.

Ayittey, George. *Africa Unchained: The Blueprint for Africa's Future*. Palgrave Macmillan US, 2016.

Azikiwe, Nnamdi. "The Future of Pan-Africanism." Nigerian High Commission, London, 1961.

Bahri, Deepika. "The Digital Diaspora: South Asians in the New Pax Electronica." In *In Diaspora: Theories, Histories, Texts*, edited by Makarand R. Paranjape. Indialog Publications, 2001.

Balogun, Oluwakemi M. "Beauty and the Bikini: Embodied Respectability in Nigerian Beauty Pageants." *African Studies Review* 62, no. 2 (June 2019): 80–102. https://doi.org/10.1017/asr.2018.125.

Bamba, Abou B. *African Miracle, African Mirage: Transnational Politics and the Paradox of Modernization in Ivory Coast*. Ohio University Press, 2016.

Bangura, Abdul Karim. *African Mathematics: From Bones to Computers*. University Press of America, 2012.

Baran, Paul. "On Distributed Communications Networks." *IEEE Transactions on Communications* 12, no. 1 (1964): 1–9.

Barassi, Veronica. *Activism on the Web: Everyday Struggles Against Digital Capitalism*. Routledge, 2017.

Barbrook, Richard, and Andy Cameron. "The Californian Ideology." *Science as Culture* 6, no. 1 (January 1996): 44–72. https://doi.org/10.1080/09505439 609526455.

Bastian, Misty. "Nationalism in a Virtual Space: Immigrant Nigerians on the Internet." *West Africa Review* 23, no. 1 (1999): 1–9.

Begazo, Tania, Moussa P. Blimpo, and Mark Andrew Dutz. *Digital Africa: Technological Transformation for Jobs*. World Bank Group, 2023.

Belayachi, Karim, Leora Klapper, and Douglas Randall. "Entrepreneurship Database 2012," *Viewpoint: Public Policy for the Private Sector* 334 (October 2012).

Benjamin, Ruha, ed. *Captivating Technology: Race, Carceral Technoscience, and Liberatory Imagination in Everyday Life*. Duke University Press, 2019.

Benkler, Yochai. *The Wealth of Networks: How Social Production Transforms Markets and Freedom*. Yale University Press, 2006.

Bernal, Victoria. *Nation as Network: Diaspora, Cyberspace, and Citizenship*. University of Chicago Press, 2014.

———. "Nationalist Networks: The Eritriean Diaspora Online." In *Diasporas in the New Media Age*, edited by A. Alonso and P. J. Oiarzabal. University of Nevada Press, 2010.

Bhabha, Homi K. *The Location of Culture*. Routledge, 1994.

Bijker, Wiebe E., Thomas Parke Hughes, and Trevor Pinch, eds. *The Social Construction of Technological Systems: New Directions in the Sociology and History of Technology*. MIT Press, 1987.

Blum, Andrew. *Tubes: A Journey to the Center of the Internet*. HarperCollins, 2014.

Blyden, Edward Wilmot. *Christianity, Islam and the Negro Race*. Black Classic Press, 1994. Originally published in 1887.

Bodomo, Adams B., Stephen Agyepong, Daniel Appiah, Samuel Asomaning, Yaw Agyaba, Samuel Aggrey, and Anthony Sallar. *The Okyeame Yearbook: A Transcript of Electronic Mail Discussions by a Group of Ghanaian Residents* Abroad. 1993. https://okyeame.net/okyeame/okyeameyearbook.pdf.

Boellerstorff, Tom. *Coming of Age in Second Life: An Anthropologist Explores the Virtually Human*. Princeton University Press, 2008.

Bowler, Anne. "Politics as Art: Italian Futurism and Fascism." *Theory and Society* (1991): 763–94.

Brinkerhoff, Jennifer M. "Digital Diasporas and Conflict Prevention: The Case of Somalinet.Com." *Review of International Studies* 32, no. 1 (January 2006): 25–47. https://doi.org/10.1017/S0260210506006917.

Brock, André. "Critical Technocultural Discourse Analysis." *New Media & Society* 20, no. 3 (2018): 1012–30.

Brown, Matthew H. *Indirect Subjects: Nollywood's Local Address*. Duke University Press, 2021.

Brubaker, Rogers. "The 'Diaspora' Diaspora." *Ethnic and Racial Studies* 28, no. 1 (2005): 1–9.

Burden-Stelly, Charisse. *Black Scare, Red Scare: Theorizing Capitalist Racism in the United States.* University of Chicago Press, 2023.

Burrell, Jenna. *Invisible Users: Youth in the Internet Cafés of Urban Ghana.* MIT Press, 2012.

Business Elites Africa. "Dangote and Elumelu Under Fire for Attempting to Corner EndSARS Protesters." October 14, 2020. https://businesselitesafrica .com/dangote-elumelu-under-fire-for-attempting-to-corner-endsars -protesters/.

Campt, Tina. "The Crowded Space of Diaspora: Intercultural Address and the Tensions of Diasporic Relation." *Radical History Review* 83 (Spring 2002): 94–113.

Candidatu, Laura, Koen Leurs, and Sandra Ponzanesi. "Digital Diasporas: Beyond the Buzzword; Toward a Relational Understanding of Mobility and Connectivity." in *The Handbook of Diasporas, Media, and Culture*, edited by Jessica Retis and Roza Tsagarousianou. Wiley Blackwell, 2019.

Cardoso, Fernando Henrique, and Enzo Faletto. *Dependency and Development in Latin America.* University of California Press, 1979.

Carmody, Pádraig Risteard. *The New Scramble for Africa.* Polity Press, 2011.

Castells, Manuel. *Networks of Outrage and Hope: Social Movements in the Internet Age*, 2nd ed. Polity Press, 2015.

———. *The Rise of the Network Society.* Blackwell Publishers, 1996.

Celina, Runako, Henry Mhango, and Chiara Francavilla. "Racism for Sale." *BBC: Africa Eye*, June 12, 2022. https://www.bbc.com/news/av/world-africa -61764466.

Cheddie, Janice. "From Slaveship to Mothership and Beyond: Thoughts on a Digital Diaspora." In *Desire by Design: Body, Territories, and New Technologies*, edited by Cutting Edge. I. B. Tauris, 1999.

Chipidza, Wallace, and Dorothy Leidner. "A Review of the ICT-Enabled Development Literature: Towards a Power Parity Theory of ICT4D." *Journal of Strategic Information Systems* 28, no. 2 (June 2019): 145–74. https://doi .org/10.1016/j.jsis.2019.01.002.

Chironga, Mutsa, Hilary De Grandis, and Yassir Zouaoui. "Mobile Financial Services in Africa: Winning the Battle for the Customer." McKinsey & Company. September 1, 2017. https://wwww.mckinsey.com/industries/financial -services/our-insights/mobile-financial-services-in-africa-winning-the-battle -for-the-customer.

Chun, Wendy Hui Kyong. "The Enduring Ephemeral, or the Future Is a Memory." *Critical Inquiry* 35, no. 1 (2008): 148–71.

Chun, Wendy Hui Kyong, and Thomas Keenan. *New Media, Old Media: A History and Theory Reader.* Psychology Press, 2006.

Clark, Gracia. *African Market Women: Seven Life Stories from Ghana.* Indiana University Press, 2010.

Clark, Msia Kibona, Phiwokuhle Mnyandu, and Loy L. Azalia, eds. *Pan African Spaces: Essays on Black Transnationalism.* Lexington Books, 2019.

Clark, Msia Kibona, and Wunpini Fatimata Mohammed, eds. *African Women in Digital Spaces: Redefining Social Movements on the Continent and in the Diaspora.* Mkuki na Nyota, 2023.

Clarke, John Henrik, and Amy Jacques Garvey. *Marcus Garvey and the Vision of Africa*. Random House, 1974.

Clarke, Kamari Maxine. "Kony 2012, the ICC, and the Problem with the Peace-and-Justice Divide." *Proceedings of the ASIL Annual Meeting* 106 (2012): 309–13. https://doi.org/10.5305/procannmeetasil.106.0309.

Clarke, Kamari Maxine, and Deborah A. Thomas, eds. *Globalization and Race: Transformations in the Cultural Production of Blackness*. Duke University Press, 2006.

Cobla, Georgina Maku, and Eric Osei-Assibey. "Mobile Money Adoption and Spending Behaviour: The Case of Students in Ghana." *International Journal of Social Economics* 45, no. 1 (January 8, 2018): 29–42. https://doi.org/10.1108/IJSE-11-2016-0302.

Coetzee, Paul, and A. P. J. Roux, eds. *Philosophy from Africa: A Text with Readings*. 2nd ed. Oxford University Press, 2002.

Cohen, Robin. *Global Diasporas: An Introduction*. University of Washington Press, 1997.

Coleman, E. Gabriella. *Coding Freedom: The Ethics and Aesthetics of Hacking*. Princeton University Press, 2013.

Cooke, Edgar, Sarah Hague, and Andy McKay. *The Ghana Poverty and Inequality Report: Using the 6th Ghana Living Standards Survey*. University of Sussex, 2016.

Cossa, José, Lesley Le Grange, and Yusef Waghid. "Cosmo-Ubuntu: Toward a New (Exterior to Modernity) Theorizing About the Human, the Cosmos, and Education." *Comparative Education Review* 64, no. 4 (2020): 753–56.

Cottrell, R. Les. "Pinging Africa: A Decadelong Quest Aims to Pinpoint the Internet Bottlenecks Holding Africa Back." *IEEE Spectrum* 50, no. 2 (2013): 54–59.

Coulibaly, Brahima S. "In Defense of the 'Africa Rising' Narrative." Brookings Institute. June 27, 2017. https://www.brookings.edu/articles/in-defense-of-the-africa-rising-narrative/.

Daily Guide. "Accra Floods: More than 100 Feared Dead after Explosion." Modern Ghana, July 4, 2015. https://www.modernghana.com/news/621226/accra-floods-more-than-100-feared-dead-after-explosion.html.

Daly, Samuel Fury Childs. "Ghana Must Go: Nativism and the Politics of Expulsion in West Africa, 1969–1985." *Past & Present* 259, no. 1 (May 2023): 229–61. https://doi.org/10.1093/pastj/gtac006.

David-West, Olayinka, Nkemdilim Iheanachor, and Immanuel Ovemeso Umukoro. "Mobile Money as a Frugal Innovation for the Bottom of the Pyramid—Cases of Selected African Countries." *Africa Journal of Management* 5, no. 3 (2019): 274–302.

Davies, Sarah R. "Characterizing Hacking: Mundane Engagement in US Hacker and Makerspaces." *Science, Technology, & Human Values* 43, no. 2 (March 2018): 171–97. https://doi.org/10.1177/0162243917703464.

Davis, Joshua Clark. *From Head Shops to Whole Foods: The Rise and Fall of Activist Entrepreneurs*. Columbia University Press, 2017.

Dean, Jodi. "Neofeudalism: The End of Capitalism?" *Los Angeles Review of Books*, May 12, 2020. https://lareviewofbooks.org/article/neofeudalism-the-end-of-capitalism/.

Decker, Corrie, and Elizabeth McMahon. *The Idea of Development in Africa.* Cambridge University Press, 2021.

Deleuze, Gilles, and Félix Guattari. *A Thousand Plateaus: Capitalism and Schizophrenia.* University of Minnesota Press, 1987.

Dery, Mark. "Black to the Future: Interviews with Samuel R. Delany, Greg Tate, and Tricia Rose." In *Flame Wars: The Discourse of Cyberculture*, edited by Mark Dery, 179–222. Duke University Press, 1994.

Dey, Bidit, and Faizan Ali. "A Critical Review of the ICT for Development Research." In *ICTs in Developing Countries: Research, Practices and Policy Implications*, edited by Bidit Dey, Karim Sorour, and Raffaele Filieri, 3–23. Palgrave Macmillan UK, 2016. https://doi.org/10.1057/9781137469502_1.

Diaspora Affairs Bureau. "Diaspora Affairs Bureau." Accessed May 1, 2014. https://diasporaaffairs.gov.gh.

Diawara, Manthia, and Silvia Kolbowski. "Homeboy Cosmopolitan." *October* 83 (Winter 1998): 51–70.

Dijk, Jan van. *The Digital Divide.* Polity, 2020.

Diminescu, Dana, and Benjamin Loveluck. "Trances of Dispersion." *Crossings: Journal of Migration & Culture* 5, no. 1 (2014): 23–39.

Donner, Jonathan. *After Access: Inclusion, Development, and a More Mobile Internet.* MIT Press, 2015.

———. "The Rules of Beeping: Exchanging Messages Via Intentional 'Missed Calls' on Mobile Phones." *Journal of Computer-Mediated Communication* 13, no. 1 (2007): 1–22. https://doi.org/10.1111/j.1083-6101.2007.00383.x.

Du Bois, W. E. B. *The World and Africa: An Inquiry into the Part Which Africa Has Played in World History.* International Publishers, 1947.

"Duapa Challenge—The Campus Duel.'" British Council. Accessed January 31, 2018. https://www.britishcouncil.org.gh/programmes/society/social-enterprise/duapa-challenge-campus-duel.

Dudek, Sam, with Kevin Bannerman-Hutchful. "Positive Effects of Mobile Money on Financial Inclusivity." *Equilibrium: Undergraduate Journal of Economics* 8 (June 8, 2018).

Dunn, Hopeton S., Dumisani Moyo, William O. Lesitaokana, and Shanade Bianca Barnabas, eds. *Re-Imagining Communication in Africa and the Caribbean: Global South Issues in Media, Culture and Technology.* Palgrave Macmillan, 2021.

Dunning, John. *Making Globalization Good: The Moral Challenges of Global Capitalism.* Oxford University Press, 2003.

Dyer, Richard. *White.* Routledge, 1997.

The Economist. "Africa Rising." December 3, 2011. https://www.economist.com/leaders/2011/12/03/africa-rising.

Eddie-Quartey, A. C. "The Story of GhanaFest-Chicago." *Modern Ghana*, August 12, 2005. https://www.modernghana.com/news/83919/the-story-of-ghanafest-chicago.html.

Edozie, Rita Kiki. *"Pan" Africa Rising: The Cultural Political Economy of Nigeria's Afri-Capitalism and South Africa's Ubuntu Business.* Contemporary African Political Economy. Palgrave Macmillan, 2017.

———. "Pan-Africanism Is Africa's Third Way: The Cultural Relevance of African Political Economy." In *The Palgrave Handbook of African Politics, Governance and Development,* edited by Samuel Ojo Oloruntoba and Toyin Falola. Palgrave Macmillan, 2018.

Edwards, Brent Hayes. *The Practice of Diaspora: Literature, Translation, and the Rise of Black Internationalism.* Harvard University Press, 2003.

Eglash, Ron, et al. *Appropriating Technology: Vernacular Science and Social Power.* University of Minnesota Press, 2004.

Ekotto, Frieda, and Kenneth W. Harrow. *Rethinking African Cultural Production.* Indiana University Press, 2015.

Elumelu, Tony. "Africapitalism: The Path to Economic Prosperity and Social Wealth." White paper, TEF. July 2022. https://www.tonyelumelufoundation .org/wp-content/uploads/dlm_uploads/2022/07/Africapitalism_The-Path-To -Economic-Prosperity-and-Social-Wealth.pdf.

Elumelu, Tony, and Heir Holdings. *The Africapitalist* 1, no. 1 (quarter 4, 2011): 1.

Ellison, Ralph. *Shadow and Act.* Vintage International, 1995. Originally published 1964.

Ellul, Jacques. *The Technological Society.* Vintage Books, 1954.

Emeagwali, Gloria, and Edward Shizha, eds. *African Indigenous Knowledge and the Sciences: Journeys into the Past and Present.* Anti-Colonial Educational Perspectives for Transformative Change, vol. 4. Brill 2016. https://doi .org/10.1007/978-94-6300-515-9.

Eshun, Kodwo. "Further Considerations on Afrofuturism." *CR: The New Centennial Review* 3, no. 2 (2003): 287–302.

Essien, Kwame. *Brazilian-African Diaspora in Ghana: The Tabom, Slavery, Dissonance of Memory, Identity, and Locating Home.* Michigan State University Press, 2016.

Evans, Martha, and Ian Glenn. "'TIA—This Is Africa': Afropessimism in Twenty-First-Century Narrative Film." *Black Camera* 2, no. 1 (2010): 14. https://doi.org/10.2979/blc.2010.2.1.14.

Everett, Anna. *Digital Diaspora: A Race for Cyberspace.* SUNY Series, Cultural Studies in Cinema/Video. SUNY Press, 2009.

Ewing, Kamahra. "Transatlantic Connections: Reception and Production of Nollywood Films in Brazil." *Journal of African Cultural Studies* 31, no. 2 (May 4, 2019): 196–211.

Falch, Morten, and Amos Anyimadu. "Tele-Centres as a Way of Achieving Universal Access—the Case of Ghana." *Telecommunications Policy* 27, nos. 1–2 (February 2003): 21–39. https://doi.org/10.1016/S0308-5961(02)00092-7.

Falcon, Ernesto. "Where Net Neutrality Is Today and What Comes Next." Electronic Frontier Foundation, December 28, 2021. https://www.eff.org /deeplinks/2021/12/where-net-neutrality-today-and-what-comes-next-2021 -review.

Fasselt, Rebecca. "'I'm Not Afropolitan—I'm of the Continent': A Conversation with Yewande Omotoso." *The Journal of Commonwealth Literature* 50, no. 2 (June 2015): 231–46. https://doi.org/10.1177/0021989414552922.

Ferguson, James. *Global Shadows: Africa in the Neoliberal World Order*. Duke University Press, 2006.

Figueroa-Vasquez, Yomaira C. *Decolonizing Diasporas: Radical Mappings of Afro-Atlantic Literatures*. Evanston: Northwestern University Press, 2021.

Fioramonti, Lorenzo. "The 'Africa Rising Story' Was Based on Faulty Logic—Here's How to Fix It." The Conversation. October 30, 2017. https://the conversation.com/the-africa-rising-story-was-based-on-faulty-logic-heres -how-to-fix-it-86327.

Fitzgibbon, Will. "The Mauritius Leaks." International Consortium of Investigative Journalists. July 23, 2019. https://www.icij.org/investigations/mauritius -leaks/treasure-island-leak-reveals-how-mauritius-siphons-tax-from-poor -nations-to-benefit-elites/.

Fortunati, Leopoldina, Raul Pertierra, and Jane Vincent. *Migration, Diaspora and Information Technology in Global Societies*. Edited by Leopoldina Fortunati, Raul Pertierra, and Jane Vincent. Routledge, 2012.

Frankema, Ewout, and Marlous van Waijenburg. "Africa Rising? A Historical Perspective." Africa Is a Country. October 17, 2018. https://africasacountry .com/2018/10/africa-rising-a-historical-perspective.

Freedom House. "Ghana." 2022. https://freedomhouse.org/country/ghana /freedom-net/2022.

———. "Transnational Repression Is a Growing Threat to Global Democracy." June 4, 2021. https://freedomhouse.org/article/new-report-transnational -repression-growing-threat-global-democracy.

Friederici, Nicolas, Michel Wahome, and Mark Graham. *Digital Entrepreneurship in Africa: How a Continent Is Escaping Silicon Valley's Long Shadow*. MIT Press, 2020.

Friedman, Thomas. *The World Is Flat: A Brief History of the Twenty-First Century*. Farrar, Straus and Giroux, 2005.

Gagliardone, Iginio. *China, Africa, and the Future of the Internet*. Zed, 2019.

Gaines, Kevin K. *American Africans in Ghana: Black Expatriates and the Civil Rights Era*. University of North Carolina Press, 2006.

Gaskins, Nettrice. "Techno-Vernacular Creativity and Innovation Across the African Diaspora and Global South." In *Captivating Technology: Race, Carceral Technoscience, and Liberatory Imagination in Everyday Life*, edited by Ruha Benjamin, 252–74. Duke University Press, 2019.

Gershenhorn, Jerry. "'Not an Academic Affair': African American Scholars and the Development of African Studies Programs in the United States, 1942– 1960." *Journal of African American History* 94, no. 1 (2009): 44–68.

Geschiere, Peter, Birgit Meyer, and Peter Pels, eds. *Readings in Modernity in Africa*. Indiana University Press, James Currey, Unisa Press, 2008.

Ghana National Council of Chicago. "Virtual Ghana Fest." Facebook, July 25, 2020. https://www.facebook.com/GNCCHICAGO/videos/3444384302 91924.

Ghana Statistical Service. *Ghana 2021 Population and Housing Census: Population of Regions and Districts*. 2021. https://statsghana.gov.gh/gssmain /fileUpload/pressrelease/2021%20PHC%20General%20Report%20Vol %203A_Population%20of%20Regions%20and%20Districts_181121.pdf.

Gifford, Paul. *Ghana's New Christianity, New Edition: Pentecostalism in a Globalising African Economy*. Indiana University Press, 2004.

Gikandi, Simon. "Foreword: On Afropolitanism." In *Negotiating Afropolitanism: Essays on Borders and Spaces in Contemporary African Literature and Folklore*, edited by Jennifer Wawrzinek and J. K. S. Makokha. Rodopi, 2011.

Gilroy, Paul. *The Black Atlantic: Modernity and Double Consciousness*. Harvard University Press, 1993.

Girancourt, Francois Jurd de, Mayowa Kuyoro, Nii Amah Ofosu-Amaah, Edem Seshie, and Frederick Twum. "How the COVID-19 Crisis May Affect Electronic Payments in Africa." McKinsey & Company, 2020. https://www .mckinsey.com/industries/financial-services/our-insights/how-the-covid-19 -crisis-may-affect-electronic-payments-in-africa.

Golumbia, David. *The Politics of Bitcoin: Software as Right-Wing Extremism*. University of Minnesota Press, 2016.

Goody, Jack. *Technology, Tradition, and the State in Africa*. Oxford University Press, 1971.

Goyal, Yogita. "When Was the Afropolitan? Thinking Literary Genealogy." *PMLA/Publications of the Modern Language Association of America* 136, no. 5 (October 2021): 778–84. https://doi.org/10.1632/S0030812921000675.

Groupe Spécial Mobile (GSMA). "The Mobile Economy: Sub-Saharan Africa 2018." 2018. http://www.gsma.com/r/mobileeconomy/.

———. *The State of Mobile Internet Connectivity 2019*. 2019. https://www .gsma.com/mobilefordevelopment/wp-content/uploads/2019/07/GSMA -State-of-Mobile-Internet-Connectivity-Report-2019.pdf.

———. *The State of Mobile Internet Connectivity 2024*. October 2024. https://www.gsma.com/r/wp-content/uploads/2024/10/The-State-of-Mobile -Internet-Connectivity-Report-2024.pdf.

———. *The State of Mobile Internet Connectivity 2021*. 2021. https://www .gsma.com/r/wp-content/uploads/2021/09/The-State-of-Mobile-Internet -Connectivity-Report-2021.pdf.

———. *The State of the Industry Report on Mobile Money 2024*. April 2024. https://www.gsma.com/sotir/wp-content/uploads/2024/03/GSMA-SOTIR -2024_Report_v7-2.pdf.

———. *Sub-Saharan Africa Mobile Economy 2013*. November 8, 2013. https:// www.gsma.com/newsroom/resources/gsma-mobile-economy-sub-saharan -africa-report-2013/.Hadjor, Kofi Buenor. *Nkrumah and Ghana: The Dilemma of Post-Colonial Power*. Kegan Paul International, 1988.

Hall, Stuart. "Notes on Deconstructing 'the Popular.'" In *Foundations of Cultural Studies*, edited by David Morley, Catherine Hall, and Bill Schwarz. Duke University Press, 2019.

———. *Representation: Cultural Representations and Signifying Practices*. Sage and Open University, 1997.

Halter, Marilyn, and Violet Showers Johnson. *African & American: West Africans in Post-Civil Rights America*. New York University Press, 2016.

Haltiwanger, John C., Ron S. Jarmin, and Javier Miranda. "Who Creates Jobs? Small vs. Large vs. Young." *SSRN Electronic Journal*, August 1, 2010. US Census Bureau Center for Economic Studies Paper No. CES-WP- 10-17. https://doi.org/10.2139/ssrn.1666157.

Harris, Joseph E. *Global Dimensions of the African Diaspora*. Howard University Press, 1982.

Hauge, Jostein. "Africa's Economic 'Rise' Does Not Reflect Reality." *The Guardian*, September 3, 2014. https://www.theguardian.com/global-development /poverty-matters/2014/sep/03/africa-economic-rise-does-not-reflect-reality.

Herskovits, Melville J. *The Myth of the Negro Past*. Harper and Bros., 1941.

Hersman, Erik. "Mobilizing Tech Entrepreneurs in Africa (Innovations Case Narrative: iHub)." *Innovations: Technology, Governance, Globalization* 7, no. 4 (October 1, 2012): 59–67. https://doi.org/10.1162/INOV_a_00152.

Hippel, Eric von. *Democratizing Innovation*. MIT Press, 2005.

Holsey, Bayo. *Routes of Remembrance: Refashioning the Slave Trade in Ghana*. University of Chicago Press, 2008.

Honeyman, Catherine A. *The Orderly Entrepreneur: Youth, Education, and Governance in Rwanda*. Anthropology of Policy. Stanford University Press, 2016.

hooks, bell. *Black Looks: Race and Representation*. South End Press, 1992.

Hopkins, Terence K., and Immanuel Maurice Wallerstein. *World-Systems Analysis: Theory and Methodology*. Explorations in the World-Economy. Sage Publications, 1982.

Horst, Heather A., and Daniel Miller. *The Cell Phone: An Anthropology of Communication*. Oxford & Berg Publishers, 2006.

Hughes, Thomas P. *Human-Built World: How to Think About Technology and Culture*. University of Chicago Press, 2004.

ICANN. "Update on AFRINIC." September 22, 2023. https://www.icann.org/en /announcements/details/icann-update-on-afrinic-22-09-2023-en.

Idemudia, Uwafiokun, and Kenneth Amaeshi, eds. *Africapitalism: Sustainable Business and Development in Africa*. Routledge, 2019.

Ighobor, Kingsley. "Closing Africa's Wealth Gap." *Africa Renewal*, December 2017–March 2018. https://www.un.org/africarenewal/magazine/december -2017-march-2018/closing-africa%E2%80%99s-wealth-gap.

International Telecommunications Union (ITU). "Digital Technologies to Achieve the UN SDGs." May 31, 2022. https://www.itu.int:443/en/mediacentre/back grounders/Pages/icts-to-achieve-the-united-nations-sustainable-development -goals.aspx.

———. *The ICT Development Index*, 2024. United Nations, June 2024. https:// www.itu.int/itu-d/reports/statistics/idi2024/.

———. "Measuring Digital Development: State of Digital Development and Trends." 2025. Reports in the Africa, Americas, Asia/Pacific and European Regions. https://www.itu.int/hub/?s=State+of+digital+development+and +trends

———. "Measuring Digital Development Facts and Figures." 2021. https:// www.itu.int/en/ITU-D/Statistics/Documents/facts/FactsFigures2021.pdf.

———. *Measuring the Information Society 2013*. 2013. https://www.itu.int /en/ITU-D/Statistics/Documents/publications/mis2013/MIS2013_without _Annex_4.pdf.

———. "World Telecommunication/ICT Indicators Database." Accessed May 2014. https://www.itu.int/en/ITU-D/Statistics/Pages/publications/wtid.aspx.

Irani, Lilly. "Hackathons and the Making of Entrepreneurial Citizenship." *Science, Technology, & Human Values* 40, no. 5 (September 2015): 799–824. https://doi.org/10.1177/0162243915578486.

Irving, Larry, et al. "Falling through the Net: A Survey of the 'Have Nots' in Rural and Urban America." National Telecommunications and Information Administration (NTIA), July 1995. https://www.ntia.gov/page/falling -through-net-survey-have-nots-rural-and-urban-america.

Jackson, John L. *Thin Description: Ethnography and the African Hebrew Israelites of Jerusalem*. Harvard University Press, 2013.

Jahajeeah, Jess Auerbach. "What a New University in Africa Is Doing to Decolonise Social Sciences." The Conversation. May 13, 2017. https://the conversation.com/what-a-new-university-in-africa-is-doing-to-decolonise -social-sciences-77181.

Jayawardane, M. Neelika. "'The Capacity-Building-Workshop-in-Africa Hokum.'" *Journal of African Cultural Studies* 31, no. 3 (September 2, 2019): 276–80. https://doi.org/10.1080/13696815.2019.1630265.

Jordan, Tim, and Paul A. Taylor. *Hacktivism and Cyberwars: Rebels with a Cause?* Routledge, 2004.

Jordans, Eva, Bettina Ng'weno, and Helen Spencer-Oatey. *Developing Global Leaders: Insights from African Case Studies*. Palgrave Macmillan, 2020. https://doi.org/10.1007/978-3-030-14606-1.

Kabwe, Mwenya B. "Transgressing Boundaries: Making Theatre from an Afropolitan Perspective." *South African Theatre Journal* 21, no. 1 (January 2007): 48. https://doi.org/10.1080/10137548.2007.9687853.

Karim, Karim H. *The Media of Diaspora*. Routledge, 2003.

Kelly, Kevin. "The Computational Metaphor." In *The New Media Theory Reader*, edited by Robert Hassan and Julian Thomas. Open University Press, 1998.

Kimble, David. *A Political History of Ghana: The Rise of Gold Coast Nationalism, 1850–1928*. Clarendon Press, 1963.

Kimeria, Ciku. "A White Founder's $1 Million Nairobi Food Startup Aims to Fix a Problem Kenyans Say Doesn't Exist." Quartz. June 25, 2021. https://qz .com/africa/2024177/white-privilege-in-the-african-startup-scene-draws-ire -of-kenyans.

Kleine, D. *Technologies of Choice? ICTs, Development, and the Capabilities Approach*. MIT Press, 2013.

Konadu, Kwasi, and Clifford C. Campbell, eds. *The Ghana Reader: History, Culture, Politics*. Duke University Press, 2016.

Kraidy, Marwan M. *Communication and Power in the Global Era: Orders and Borders*. Routledge, 2013.

Kreamer, Christine Mullen, Erin Haney, Katharine Monsted, Karel Nel, and Randall Bird. *African Cosmos: Stellar Arts*. National Museum of African Art, Smithsonian Institution; Monacelli Press, 2012.

Kuo, Lily. "Africa Wasn't 'Rising' Before and It's Not 'Reeling' Now." Quartz Africa. October 2016. https://qz.com/africa/816006/africa-wasnt-rising-before -and-its-not-reeling-now.

Kvasny, Lynette. "The Role of the Habitus in Shaping Discourses about the Digital Divide." *Journal of Computer-Mediated Communication* 10, no. 2 (January 1, 2005). https://doi.org/10.1111/j.1083-6101.2005.tb00242.x.

Kwami, Janet D. "Development from the Margins? Mobile Technologies, Transnational Mobilities, and Livelihood Practices Among Ghanaian Women Traders." *Communication, Culture & Critique* 9, no. 1 (2016): 148–68. https://doi.org/10.1111/cccr.12136.

Laguerre, Michel. *Diaspora, Politics, and Globalization*. Palgrave Macmillan, 2006.

———. *Urban Multiculturalism and Globalization in New York City: An Analysis of Diasporic Temporalities*. Palgrave Macmillan, 2003.

Laguerre, Michel S. *The Multisited Nation: Crossborder Organizations, Transfrontier Infrastructure, and Global Digital Public Sphere*. Palgrave Macmillan, 2016.

Larkin, Brian. *Signal and Noise: Media, Infrastructure, and Urban Culture in Nigeria*. Duke University Press, 2008.

Latour, Bruno. *Reassembling the Social: An Introduction to Actor-Network Theory*. Oxford University Press, 2005.

———. "Technology Is Society Made Durable." *The Sociological Review* 38, no. 1 (May 1, 1990): 103–31. https://doi.org/10.1111/j.1467-954X.1990.tb 03350.x.

Latour, Bruno, and Steve Woolgar. *Laboratory Life: The Construction of Scientific Facts*. Princeton University Press, 1986.

Law, John, ed. *A Sociology of Monsters: Essays on Power, Technology, and Domination*. Routledge, 1991.

Leiner, Barry, et al. "A Brief History of the Internet." The Internet Society, 1997. https://www.internetsociety.org/internet/history-internet/brief-history-internet/.

Lemonnier, Pierre. *Elements for an Anthropology of Technology*. University of Michigan Press, 1992.

Lessig, Lawrence. *The Future of Ideas: The Fate of the Commons in a Connected World*. Random House, 2001.

Leurs, Koen, and Kevin Smets. "Five Questions for Digital Migration Studies: Learning from Digital Connectivity and Forced Migration In(to) Europe." *Social Media + Society* 4, no. 1 (January 2018): 1–16.

Lévi-Strauss, Claude. *The Savage Mind*. University of Chicago Press, 1966.

Lévy, Pierre. *Cyberculture*. University of Minnesota Press, 2001.

Levy, Steven. *Hackers: Heroes of the Computer Revolution*. Doubleday, 1984.

MacKenzie, Donald A., and Judy Wajcman. *The Social Shaping of Technology: How the Refrigerator Got Its Hum*. Open University Press, 1985.

Madianou, Mirca, and Daniel Miller. *Migration and New Media: Transnational Families and Polymedia*. Routledge, 2012.

Mahama, John Dramani. *My First Coup d'Etat: Memories from the Lost Decades of Africa.* A&C Black, 2012.

Manovich, Lev. *The Language of New Media.* MIT Press, 2002.

Manuh, Takyiwaa, ed. *At Home in the World?: International Migration and Development in Contemporary Ghana and West Africa.* Sub-Saharan Publishers, 2005.

———. "'Efie' or the Meanings of 'Home' Among Female and Male Ghanaian Migrants in Toronto, Canada and Returned Migrants to Ghana." In *New African Diasporas*, edited by Khalid Koser. Taylor & Francis, 2003.

Marx, Leo. "Technology: The Emergence of a Hazardous Concept." In *Technology and the Rest of Culture*, edited by Arien Mack. Ohio State University Press, 1997.

Mathew, Ashwin J. "The Myth of the Decentralised Internet." *Internet Policy Review* 5, no. 3 (September 30, 2016). https://doi.org/10.14763/2016.3.425.

Matlon, Jordanna. *A Man Among Other Men: The Crisis of Black Masculinity in Racial Capitalism.* Cornell University Press, 2022.

Mavhunga, Clapperton Chakanetsa, ed. *What Do Science, Technology, and Innovation Mean from Africa?* MIT Press, 2017.

Mbembe, Achille. "Afropolitanism." In *Africa Remix: Contemporary Art of a Continent*, edited by Njami Simon and Lucy Durán. Jacana Media, 2007.

———. *On the Postcolony.* University of California Press, 2001.

Media Ownership Monitor Ghana. "Bellaart Investments B.V. (GhanaWeb)." Accessed July 3, 2021. https://ghana.mom-gmr.org/en/about/faq/.

Melber, Henning, ed. *The Rise of Africa's Middle Class: Myths, Realities and Critical Engagements.* Zed Books, 2016.

Meredith, Martin. *The Fate of Africa: From the Hopes of Freedom to the Heart of Despair; a History of Fifty Years of Independence.* Public Affairs, 2005.

Miescher, Stephan. "The Akosombo Dam and the Quest for Rural Electrification in Ghana." In *Electric Worlds/Mondes Électriques: Creations, Circulations, Tensions, Transitions (19th–21st C.)*, edited by Alain Beltran, Léonard Laborie, Pierre Lanthier, and Stéphanie Le Gallic. Peter Lang, 2016.

———. *A Dam for Africa: Akosombo Stories from Ghana.* Indiana University Press, 2022.

Miescher, Stephan, and Leslie Ashbaugh. "Been-To Visions: Transnational Linkages Among a Ghanaian Dispersed Community in the Twentieth Century." *Ghana Studies Journal* 2 (1999): 5–95.

Migration Policy Institute. "The Ghanaian Diaspora in the United States." Rockefeller Aspen Diaspora Program, 2015. https://www.migrationpolicy.org/sites/default/files/publications/RAD-Ghana.pdf.

Miller, Daniel, and Don Slater. *The Internet: An Ethnographic Approach.* Berg Publishers, 2000.

Mirvis, Philip, and Bradley Googins. "Catalyzing Social Entrepreneurship in Africa: Roles for Western Universities, NGOs and Corporations." *Africa Journal of Management* 4, no. 1 (January 2, 2018): 57–83. https://doi.org/10.1080/23322373.2018.1428020.

Mkandawire, Thandika, ed. *African Intellectuals: Rethinking Politics, Language, Gender, and Development.* Zed Books, 2005.

———. "Thinking about Developmental States in Africa." *Cambridge Journal of Economics* 25, no. 3 (May 1, 2001): 289–314. https://doi.org/10.1093/cje /25.3.289.

Mohammed-Nurudeen, Mahmud. "Ejura Shooting: 2 Families Receive ¢500k Compensation from Government." MyJoyOnline. February 6, 2022. https:// www.myjoyonline.com/ejura-shooting-2-families-receive-%c2%a2500k -compensation-from-government/.

Mosse, George L. "The Political Culture of Italian Futurism: A General Perspective." *Journal of Contemporary History* 25, no. 2 (1990): 253–68.

Moyo, Dambisa. *Dead Aid: Why Aid Is Not Working and How There Is a Better Way for Africa.* Farrar, Straus and Giroux, 2009.

Mudhai, Okoth Fred. *Civic Engagement, Digital Networks, and Political Reform in Africa.* Palgrave-MacMillan Press, 2013.

Mudhai, Okoth Fred, Wisdom Tettey, and Franklin Banda. *African Media and the Digital Public Sphere.* Palgrave Macmillan, 2009.

Mutsvairo, Bruce. *Digital Activism in the Social Media Era: Critical Reflections on Emerging Trends in Sub-Saharan Africa.* Springer, 2016.

MyJoyOnline. "Annoh-Dompreh Apologises for #FixYourself Comment." May 4, 2021. https://www.myjoyonline.com/annoh-dompreh-apologises-for -fixyourself-comment/.

Nardi, Bonnie. *My Life as a Night Elf Priest: An Anthropological Account of World of Warcraft.* University of Michigan Press, 2010.

Nardi, Bonnie A., and Vicki O'Day. *Information Ecologies: Using Technology with Heart.* MIT Press 1999.

Ndemo, Bitange, and Tim Weiss, eds. *Digital Kenya: An Entrepreneurial Revolution in the Making.* Springer, 2016.

Neal, Mark Anthony. *Looking for Leroy: Illegible Black Masculinities.* New York University Press, 2013.

New Patriotic Party (NPP). "Manifesto 2020: Leadership of Service." Accessed December 9, 2020. https://media.peacefmonline.com/docs/202008/353419 888_327383.pdf.

Nicolás, Valenzuela-Levi. "The Written and Unwritten Rules of Internet Exclusion: Inequality, Institutions and Network Disadvantage in Cities of the Global South." *Communication & Society* 24, no. 11 (August 18, 2021): 1568–85.

Nimako, Kwame. *Economic Change and Political Conflict in Ghana, 1600–1990.* Amsterdam: Thesis Publishers, 1991.

Nkrumah, Kwame. "African Socialism Revisited." In *Africa: National and Social Revolution.* Peace and Socialism Publishers, 1967. https://www.marxists.org /subject/africa/nkrumah/1967/african-socialism-revisited.htm.

———. *Ghana: The Autobiography of Kwame Nkrumah.* Thomas Nelson and Sons, 1957.

———. *Towards Colonial Freedom: Africa in the Struggle Against World Imperialism.* Panaf, 1973.

Noble, Safiya Umoja. *Algorithms of Oppression: How Search Engines Reinforce Racism.* New York University Press, 2018.

Nyamnjoh, Francis B. *#RhodesMustFall: Nibbling at Resilient Colonialism in South Africa.* Langaa Research & Publishing CIG, 2016.

Nyerere, Julius K. "A United States of Africa." *The Journal of Modern African Studies* 1, no. 1 (March 1963): 1–6.

Obeng-Odoom, Franklin. "Africa: On the Rise, but to Where?" *Forum for Social Economics* 44, no. 3 (September 2, 2015): 234–50. https://doi.org/10.1080/07360932.2014.955040.

Ochonu, Moses E., ed. *Entrepreneurship in Africa: A Historical Approach*. Indiana University Press, 2018.

Okpewho, Isidore, and Nkiru Nzegwu, eds. *The New African Diaspora*. Indiana University Press, 2009.

Olaniyan, Tejumola. "African Cultural Studies: Of Travels, Accents, and Epistemologies." In *Rethinking African Cultural Production*, edited by Frieda Ekotto and Kenneth W. Harrow. Indiana University Press, 2015.

Olaniyan, Tejumola, and James Sweet, eds., *The African Diaspora and the Disciplines*. Indiana University Press, 2010.

Olopade, Dayo. *The Bright Continent: Breaking Rules & Making Change in Modern Africa*. HMH, 2014.

Olurounbi, Ruth. "Nigeria Central Bank Orders Cryptocurrency Accounts to Close." Bloomberg, February 5, 2021. https://www.bloomberg.com/news/articles/2021-02-05/nigerian-central-bank-orders-closure-of-cryptocurrency-accounts.

Oloruntoba, Samuel Ojo, and Toyin Falola. *The Palgrave Handbook of African Politics, Governance and Development*. Palgrave Macmillan, 2018.

Oluwole, Victor. "Top 10 African Countries with the Most Electronic Waste." August 29, 2023. https://africa.businessinsider.com/local/lifestyle/top-10-african-countries-with-the-most-electronic-waste/8czptln.

O'Neill, Kevin Lewis. "Disenfranchised: Mapping Red Zones in Guatemala City." *Environment and Planning A: Economy and Space* 51, no. 3 (May 2019): 654–69. https://doi.org/10.1177/0308518X18800069.

Onukwue, Alexander. "The Mobile Money Industry Processed More than $1 Trillion in 2021." *Quartz*, March 31, 2022. https://qz.com/africa/2149015/the-mobile-money-industry-processed-more-than-1-trillion-in-2021.

Opoku-Agyemang, Kwabena. "Lost/Gained in Translation: Oware 3D, Ananse; The Origin and Questions of Hegemony." *Journal of Gaming & Virtual Worlds* 7, no. 2 (June 1, 2015): 155–68. https://doi.org/10.1386/jgvw.7.2.155_1.

Osei, Robert Darko, Richmond Atta-Ankomah, and Monica Lambon-Quayefio. *Structural Transformation and Inclusive Growth in Ghana*. 37th ed. United Nations University-WIDER, 2020. https://doi.org/10.35188/UNU-WIDER/2020/794-1.

Osei-Opare, Nana. "Ghana and Nkrumah Revisited: Lenin, State Capitalism, and Black Marxist Orbits." *Comparative Studies in Society and History* 65, no. 2 (April 2023): 399–421. https://doi.org/10.1017/S0010417522000548.

Osseo-Asare, Abena Dove. *Atomic Junction: Nuclear Power in Africa after Independence*. Cambridge University Press, 2019.

——— . *Bitter Roots: The Search for Healing Plants in Africa*. University of Chicago Press, 2014.

Osumare, Halifu. *Hiplife in Ghana: West African Indigenization of Hip-Hop*. Palgrave Macmillan, 2013.

Otu, Kwame Edwin. "When the Lagoons Remember: An Afroqueer Futurist Reading of 'Blue Ecologies of Agitation.'" In "Gender and Sexuality in African Futurism," edited by Jacqueline-Bethel Tchouta Mougoué. Special issue, *Feminist Africa* 2, no. 2 (2021): 29–46.

Overå, Ragnhild. "Networks, Distance, and Trust: Telecommunications Development and Changing Trading Practices in Ghana." *World Development* 34, no. 7 (July 2006): 1301–15. https://doi.org/10.1016/j.worlddev.2005.11.015.

Owusu-Ankomah, P. "Emigration from Ghana: A Motor or Brake for Development." Presented at the 39th Session of the Commission on Population and Development, April 4, 2006. http://www.un.org/esa/population/cpd /cpd2006/CPD2006_Owusu_Ankomah_Statement.pdf.

Owusu-Ansah, David. *Historical Dictionary of Ghana*. 4th ed. Historical Dictionaries of Africa. Rowman & Littlefield, 2014.

Padmore, George. *Pan-Africanism or Communism: The Coming Struggle for Africa*. Doubleday, 1971. First published 1956.

Palmer, Daniel. "Mauritius to License Crypto Custodians Starting in March." February 8, 2019. https://www.coindesk.com/markets/2019/02/08/mauritius -to-license-crypto-custodians-starting-in-march/.

Penteriani, Gaia. *The e-Levy in Ghana: Economic Impact Assessment*. GSMA, March 2023. https://www.gsma.com/publicpolicy/wp-content/uploads/2023 /03/E-Levy-Ghana-Economic-Impact-Assessment.pdf.

Pfaffenberger, Bryan. "Social Anthropology of Technology." *Annual Review of Anthropology* 21 (1992): 491–516.

Pierre, Jemima. "'The Beacon of Hope for the Black Race': State Race-Craft and Identity Formation in Modern Ghana." *Cultural Dynamics* 21, no. 1 (2009): 29–50.

———. "Beyond Heritage Tourism." *Social Text* 27, no. 1 (2009): 59–81.

———. *The Predicament of Blackness: Postcolonial Ghana and the Politics of Race*. University of Chicago Press, 2013.

———. "The Racial Vernaculars of Development: A View from West Africa." *American Anthropologist* 122, no. 1 (March 2020): 86–98.

Pilling, David. "How Covid-19 Interrupted the Mauritian Economic Miracle." *Financial Times*, April 1, 2022. https://www.ft.com/content/224ab32e-f84a -4eec-a57b-da73e841d010.

Pinto, Samantha. *Difficult Diasporas: The Transnational Feminist Aesthetic of the Black Atlantic*. New York University Press, 2013.

Poppendieck, Mary, and Tom Poppendieck. *Lean Software Development: An Agile Toolkit*. Addison-Wesley, 2003.

Porter, Kiesha. "Dropifi Takes on Silicon Valley." *CNN*, July 10, 2013. http:// www.cnn.com/2013/07/10/tech/web/ghana-dropifi-silicon-valley/.

Pype, Katrien. "Branhamist Kindoki: Ethnographic Notes on Connectivity, Technology, and Urban Witchcraft in Contemporary Kinshasa." In *Pentecostalism and Witchcraft*, edited by Knut Rio, Michelle MacCarthy, and Ruy Blanes. Springer, 2017. https://doi.org/10.1007/978-3-319-56068-7_5.

Quayson, Ato. *Calibrations: Reading for the Social*. University of Minnesota Press, 2003.

Rabaka, Reiland, ed. *Routledge Handbook of Pan-Africanism*. Routledge, 2020.

Raley, Rita. *Tactical Media*. University of Minnesota Press, 2009.

Ramtohul, Ramola, and Thomas Hylland Eriksen. *The Mauritian Paradox: Fifty Years of Development, Diversity and Democracy*. Baltimore, MD: Project Muse/University of Mauritius Press, 2018.

Rao, Madanmohan. *Mobile Africa Report 2011: Regional Hubs of Excellence*. Mobile Monday, November 1, 2011. http://www.mobilemonday.net/reports/MobileAfrica_2011.pdf.

Rattray, Robert Sutherland. *The Tribes of the Ashanti Hinterland*. Vol. 2, *Ashanti*. Oxford: Clarendon Press, 1923.

Reporters Without Borders. "Mauritius." Accessed October 24, 2024. https://rsf.org/en/country/mauritius.

Rheingold, Howard. *The Virtual Community: Homesteading on the Electronic Frontier*. Addison-Wesley, 1993.

Rist, Gilbert. *The History of Development: From Western Origins to Global Faith*. 4th ed. Zed Books, 2014.

Robinson, Laura. "A Taste for the Necessary." *Information, Communication & Society* 12 (2009): 488–507.

Rodney, Walter. *How Europe Underdeveloped Africa*. Howard University Press, 1981.

Rogers, Everett M. *Diffusion of Innovations*. Free Press of Glencoe, 2005. Originally published in 1962.

Rottenburg, Richard. *Far-Fetched Facts a Parable of Development Aid*. MIT Press, 2009.

Royston, Reginold A. "At Home, Online: Affective-Exchange in Ghanaian Internet Video." In *Migrating the Black Body: Visual Culture and Diaspora*, edited by L. Raiford and H. Raphael-Hernandez. Seattle: University of Washington Press, 2017.

———. "Exploring Mobile Money in Accra 2017." Unpublished ms., African Global Ties project, n.d. https://www. raroyston.com

———. "From the Grammys to Agbogbloshie: African Viral Dance's Troubled Circuits." *The Black Scholar* 54, no. 3 (July 2, 2024): 11–18. https://doi.org/10.1080/00064246.2024.2364570.

———. "Podcasts and New Orality in the African Mediascape." *New Media & Society* 25, no. 9 (September 2023): 2455–74. https://doi.org/10.1177/14614448211021032.

———. "Reassembling Ghana: Diaspora and Innovation in the African Mediascape." PhD dissertation, University of California, Berkeley, 2014.

———. "Soulcraft: Theorizing Black Techne in African and American Viral Dance." *Social Media + Society* 8, no. 2 (April 2022). https://doi.org/10.1177/20563051221107644.

Royston, Reginold, and Krystal Strong. "Reterritorializing Twitter: African Moments 2010–2015." In *#identity: Hashtagging Race, Gender, Sexuality, and Nation*, edited by Abigail De Kosnik and Keith Feldman. University of Michigan Press, 2019.

Safran, William. "Diasporas in Modern Societies: Myths of the Homeland and Return." *Diaspora: A Journal of Transnational Studies* 1, no. 1 (1991): 83–99.

Salami, Minna. "Defining the Afropolitan." October 11, 2011. https://web.archive
.org/web/20130513142910/http://www.ariselive.com/articles/defining-the
-afropolitan/95788/.
Sangweni, Yolanda. "Ghanaian Writer Nicole Amarteifio on Creating the Hit
Web Series, 'An African City.'" AfriPOP! April 27, 2014. https://web.archive
.org/web/20140414232158/http://afripopmag.com/2014/04/ghanaian
-writer-nicole-amarteio-on-creating-the-hit-web-series-an-african-city/.
Santana, Stephanie. "Exorcizing Afropolitanism: Binyavanga Wainaina Explains
Why 'I Am a Pan-Africanist, Not an Afropolitan' at ASAUK 2012." Africa in
Words, February 8, 2013. https://africainwords.com/2013/02/08/exorcizing
-afropolitanism-binyavanga-wainaina-explains-why-i-am-a-pan-africanist
-not-an-afropolitan-at-asauk-2012/.
Schaefer, John. "Discussion Lists and Public Policy on iGhana." In *Native on the
Net: Indigenous and Diasporic Peoples in the Virtual Age*, edited by K. Land-
zelius. Routledge, 2006.
Schelenz, Laura, and Maria Pawelec. "Information and Communication Tech-
nologies for Development (ICT4D) Critique." *Information Technology for
Development* 28, no. 1 (January 2, 2022): 165–88. https://doi.org/10.1080
/02681102.2021.1937473.
Scholvin, Sören, and Moritz Breul. "An Unexpected Gateway: The Particulari-
ties of Mauritius as a Hub in Oil and Gas Gpns." *Development Southern
Africa* 38, no. 1 (2021): 139–52.
Schradie, Jen. "The Trend of Class, Race, and Ethnicity in Social Media Inequal-
ity." *Information, Communication & Society Information, Communication
& Society* 15, no. 4 (2012): 555–71.
Schwab, Klaus. *The Fourth Industrial Revolution*. Crown Business, 2017.
Sen, Amartya. *Development as Freedom*. Alfred Knopf, 1999.
Senyo, William, and Victor K. Ofoegbu. Interview with author, November 29,
2016.
Sey, Araba. "'We Use It Different, Different': Making Sense of Trends in Mobil
Phone Use in Ghana." *New Media & Society* 13, no. 3 (2011): 375–90.
Shanahan, Murray. *The Technological Singularity*. MIT Press, 2015.
Sheehy, Benedict. "Defining CSR: Problems and Solutions." *Journal of Business
Ethics* 131, no. 3 (October 2015): 625–48. https://doi.org/10.1007/s10551
-014-2281-x.
"Sheila O." WPX Power 92. Accessed April 19, 2022. https://www.power92
chicago.com/show/sheila-o/.
Shipley, Jesse Weaver. "The Market Decides If We Are Free." Africa Is a Country.
January 16, 2017. https://africasacountry.com/2017/01/the-market-decides
-if-we-are-free.
———. "Transnational Circulation and Digital Fatigue in Ghana's Azonto
Dance Craze." *American Ethnologist* 40 (2013): 362–81.
Simone, A. M. *For the City Yet to Come: Changing African Life in Four Cities*.
Duke University Press, 2004.
Slater, Don, and Janet D. Kwami. "Embeddedness and Escape: Internet and
Mobile Use as Poverty Reduction Strategies in Ghana." 2005. Information

Society Research Group Working Paper 4. https://www.researchgate.net/publication/228635823.

Smith, James H. "Tantalus in the Digital Age: Coltan Ore, Temporal Dispossession, and 'Movement' in the Eastern Democratic Republic of the Congo." *American Ethnologist* 38, no. 1 (2011): 17–35.

Southall, Roger. *The New Black Middle Class in South Africa.* James Currey, 2016.

Spencer, Amy. *DIY: The Rise of Lo-Fi Culture.* Marion Boyars, 2008.

Strong, Krystal. "Do African Lives Matter to Black Lives Matter? Youth Uprisings and the Borders of Solidarity." *Urban Education* 53, no. 2 (February 2018): 265–85. https://doi.org/10.1177/0042085917747097.

Strong, Krystal, and Christiana Kallon Kelly. "Youth Leadership for Development: Contradictions of Africa's Growing Leadership Pipeline." *The Journal of Modern African Studies* 60, no. 2 (June 2022): 217–38. https://doi.org/10.1017/S0022278X22000064.

Suderman, Alan, Frank Bajak, and Rodney Muhumuza. "Africa Internet Riches Plundered, Contested by China Broker." *San Jose Mercury News*, September 30, 2021. https://www.mercurynews.com/2021/09/30/africa-internet-riches-plundered-contested-by-china-broker/.

Suliman, Adela. "'Africans Must Move Beyond Aid and Not Be 'Beggars of the World.'"

Tawia, Eyram. *Uncompromising Passion: The Humble Beginnings of an African Video Game Industry.* BKC Consulting, 2016.

Taylor, Linnet, Gargi Sharma, Aaron Martin, and Shazade Jameson, eds. *Data Justice and Covid-19: Global Perspectives.* Meatspace Press, 2020. https://globaldatajustice.org/data-justice-and-covid-19/.

Taylor, Steven J. L. *Exiles, Entrepreneurs, and Educators: African Americans in Ghana.* SUNY Press, 2019.

Telegeography. *The State of the Network.* Telegeography, 2022. https://www2.telegeography.com/download-state-of-the-network.

Thiam, Thierno, and Gilbert Rochon. *Sustainability, Emerging Technologies, and Pan-Africanism.* Palgrave Macmillan, 2020. https://doi.org/10.1007/978-3-030-22180-5.

Thomas, Douglas. *Hacker Culture.* University of Minnesota Press, 2002.

Thompson, Robert Farris. *Flash of the Spirit: African and Afro-American Art and Philosophy.* Random House, 1983.

Thorat, Dhanashree. "Colonial Topographies of Internet Infrastructure: The Sedimented and Linked Networks of the Telegraph and Submarine Fiber Optic Internet." In *South Asian Digital Humanities: Postcolonial Mediations across Technology's Cultural Canon,* edited by Roopika Risam and Rahul K. Gairola. Routledge, Taylor & Francis Group, 2021.

Toffler, Alvin, ed. *The Futurists.* Random House 1972.

Tölöyan, Khachig. 'The Nation-State and its Others: In Lieu of a Preface." *Diaspora: A Journal of Transnational Studies* 1, no. 1(1991): 3–7.

Tomaselli, Keyan G., and Handel Kashope Wright, eds. *Africa, Cultural Studies and Difference.* Routledge, 2011.

Tufekci, Zeynep. *Twitter and Tear Gas: The Power and Fragility of Networked Protest*. Yale University Press, 2017.

Turner, Fred. *From Counterculture to Cyberculture: Stewart Brand, the Whole Earth Network, and the Rise of Digital Utopianism*. University of Chicago Press, 2006.

Turner, Lorenzo Dow. *Africanisms in the Gullah Dialect*. University of Chicago Press, 1949.

Tutton, Mark. "Young, Urban and Culturally Savvy, Meet the Afropolitans." *CNN*, February 17, 2012. http://edition.cnn.com/2012/02/17/world/africa /who-are-afropolitans/index.html, Accessed Feb 23, 2012.

Tynes, Robert. "Nation-building and the Diaspora on Leonenet: A Case of Sierra Leone in Cyberspace." *New Media & Society* 9, no. 3 (2007): 497–518.

Umoren, Imaobong Denis. *Race Women Internationalists: Activist-Intellectuals and Global Freedom Struggles*. University of California Press, 2018.

United Nations. *The Digital Economy Report 2024: Shaping an Environmentally Sustainable and Inclusive Digital Future*. United Nations, 2024. https:// unctad.org/system/files/official-document/der2024_en.pdf.

———. "Fintech for the Unbanked." Department of Economic and Social Affairs, Sustainable Development. Accessed April 12, 2023. https://sdgs .un.org/partnerships/fintech-unbanked-promoting-financial-inclusion-poor -automobile-ownership-and-building.

———. *State of Broadband Report 2021*. ITU and UNESCO, 2022.

———. *State of Broadband Report: People Centered Approaches for Universal Broadband*. ITU and UNESCO, 2021.

———. "Sustainable Development Goals." Department of Economic and Social Affairs Sustainable Development Accessed March 6, 2016. https://sdgs.un .org/goals.

———. *2021 Annual Report: Reducing Inequality*. UN Conference on Trade and Development, 2021. https://unctad.org/annual-report-2021.

———. "United Nations Sustainable Development Goals." 2011. https://sdgs .un.org/goals.

US Bureau of the Census. "American Community Survey." 2021. https://data .census.gov/table/ACSDT5Y2020.B05006.

———. "American Community Survey 1-Year Data (2005–2022)." Accessed September 1, 2024. https://www.census.gov/data/developers/data-sets/acs-1year .html

US Department of State. "Background Notes: Ghana Economy." Accessed September 23, 2011. https://www.state.gov/u-s-relations-with-ghana/#history.

Valenzuela-Levi, Nicolás. "The Written and Unwritten Rules of Internet Exclusion: Inequality, Institutions and Network Disadvantage in Cities of the Global South." *Communication & Society* 24, no. 11 (August 18, 2021): 1568–85.

Vertovec, Steven, and Robin Cohen, eds. *Migration, Diasporas, and Transnationalism*. Edward Elgar, 1999.

Villa-Vicencio, Charles, Erik Doxtader, and Ebrahim Moosa. *The African Renaissance and the Afro-Arab Spring: A Season of Rebirth?* Georgetown University Press, 2015.

Von Hesse, Hermann, and Larry W. Yarak. "A Tale of Two 'Returnee' Communities in the Gold Coast and Ghana: Accra's Tabon and Elmina's Ex-Soldiers, 1830s to the Present." *International Journal of African Historical Studies* 51, no. 2 (2018): 197–217.

Wallerstein, Immanuel M. "Africa in a Capitalist World." *Issue: A Journal of Opinion* 10, nos. 1, 2 (1980): 21–31.

Walters, Ronald W. *Pan Africanism in the Africa Diaspora: An Analysis of Modern Afrocentric Political Movements.* Wayne State, 1997.

Warf, Barney. *Global Geographies of the Internet.* Springer Netherlands, 2013.

Wark, McKenzie. *A Hacker Manifesto.* Harvard University Press, 2022.

Watkins, S. Craig. *The Young and the Digital: What the Migration to Social-Network Sites, Games, and Anytime, Anywhere Media Means for Our Future.* Beacon Press, 2009.

Wawrzinek, Jennifer. "Afropolitanism and the Novel: Mapping Material Networks in Recent Fiction from the African Diaspora." In *New Approaches to the Twenty-First-Century Anglophone Novel*, edited by Sibylle Baumbach and Birgit Neumann. Springer International Publishing, 2019. https://doi.org/10.1007/978-3-030-32598-5_13.

Weber, Max. *Economy and Society: An Outline of Interpretive Sociology.* University of California Press, 1978.

Wehr, Kevin. *DIY: The Search for Control and Self-Reliance in the 21st Century.* Routledge, 2016.

Wengraf, Lee. *Extracting Profit: Imperialism, Neoliberalism and the New Scramble for Africa.* Haymarket Books, 2018.

White, Carmen M. "Living in Zion: Rastafarian Repatriates in Ghana, West Africa." *Journal of Black Studies* 37, no. 5 (May 2007): 677–709. https://doi.org/10.1177/0021934705282379.

Whitfield, Lindsey. *Economies After Colonialism: Ghana and the Struggle for Power.* Cambridge University Press, 2020.

Wilks, Ivor. *Forests of Gold: Essays on the Akan and the Kingdom of Asante.* Ohio University Press, 1993.

Willems, Wendy. "Beyond Platform-Centrism and Digital Universalism: The Relational Affordances of Mobile Social Media Publics." *Information, Communication & Society* 24, no.12 (2021): 1677–93.

Williams, Justin. *Pan-Africanism in Ghana: African Socialism, Neoliberalism, and Globalization.* Carolina Academic Press, 2016.

Williams, Rosalind. "Afterword: An Historian's View on the Network Society." In *The Network Society: A Cross-Cultural Perspective*, edited by Manuel Castells. Edward Elgar, 2004.

Wilson, Ernest J., and Kevin R. Wong. *Negotiating the Net in Africa: The Politics of Internet Diffusion.* Lynne Rienner, 2007.

Wisnioski, Matthew H. *Engineers for Change: Competing Visions of Technology in 1960s America.* Engineering Studies Series. MIT Press, 2012.

Woo-Cumings, Meredith, ed. *The Developmental State.* Cornell University Press, 2019.

Woolgar, Steve. "Configuring the User: The Case of Usability Trials." In *A Sociology of Monsters: Essays on Power, Technology, and Domination*, edited by John Law. Routledge, 1991.

World Bank. "Archives: Ghana." Originally accessed May 1, 2018. http://www.worldbank.org/en/country/ghana/projects/all. (This web page is no longer valid; all information available on April 25, 2025, at https://countryhistorical profiles.worldbank.org/?country=GHA&searchprofile=true).

———. "Digital Economy for Africa Initiative." Accessed April 20, 2024. https://www.worldbank.org/en/programs/all-africa-digital-transformation.

———. *Ghana: World Development Indicators: 2014*. 2014. https://elibrary.worldbank.org/doi/abs/10.1596/978-1-4648-0163-1.

———. "Ghana Looks to Retool Its Economy as It Reaches Middle-Income Status." July 2011. https://www.worldbank.org/en/news/feature/2011/07/18/ghana-looks-to-retool-its-economy-as-it-reaches-middle-income-status.

Wright, Joanna. "Translating the 'Afropolitan.'" *The Media Online*, April 30, 2013. http://themediaonline.co.za/2013/04/translating-the-afropolitan/.

Yeboah, Ian E. A. *Black African Neo-Diaspora: Ghanaian Immigrant Experiences in the Greater Cincinnati, Ohio Area*. Lexington Books, 2008.

Yékú, James. *Cultural Netizenship: Social Media, Popular Culture, and Performance in Nigeria*. Indiana University Press, 2022.

Young, Crawford. *The Postcolonial State in Africa: Fifty Years of Independence, 1960–2010*. Africa and the Diaspora: History, Politics, Culture. University of Wisconsin Press, 2012.

Young African Leaders Initiative. "The Resilient Entrepreneur." US State Department. Accessed August 1, 2024. https://yali.state.gov/courses/course-5170/#/lesson/lesson-1-planning-for-resilience.

Zachary, G. Pascal. "Black Star: Ghana, Information Technology and Development in Africa." *First Monday* 9, no. 3 (March 2004). https://firstmonday.org/ojs/index.php/fm/article/download/1126/1046/9786.

Zayani, Mohamed. *Networked Publics and Digital Contention: The Politics of Everyday Life in Tunisia*. Oxford University Press, 2015.

Zeleza, Paul T. "African Diasporas: Towards a Global History." *African Studies Review* 53, no. 1 (2020): 1–19.

Index

Figures and maps are indicated by *fig.* and *map* following the page number, respectively. Page numbers followed by n indicate notes.

Founded in 1893,
UNIVERSITY OF CALIFORNIA PRESS
publishes bold, progressive books and journals
on topics in the arts, humanities, social sciences,
and natural sciences—with a focus on social
justice issues—that inspire thought and action
among readers worldwide.

The UC PRESS FOUNDATION
raises funds to uphold the press's vital role
as an independent, nonprofit publisher, and
receives philanthropic support from a wide
range of individuals and institutions—and from
committed readers like you. To learn more, visit
ucpress.edu/supportus.

www.ingramcontent.com/pod-product-compliance
Lightning Source LLC
Chambersburg PA
CBHW020830270326
41928CB00006B/483